GUICE TO

The Blue Planet

An Introduction to Earth System Science

Brian J. Skinner
Yale University

Stephen C. Porter
University of Washington

LABORATORY MANUAL

Marcia G. Bjørnerud
John M. Hughes
A. Dwight Baldwin, Jr.
Miami University—Oxford, Ohio

John Wiley & Sons, Inc.

New York   Chichester   Brisbane   Toronto   Singapore

FIGURE CREDITS

Figure 1-2: Courtesy NASA/Jet Propulsion Laboratory

Figure 1-3: Lick Observatory photograph #L5. Used with permission.

Figure 1-4: Courtesy D. Roddy, US Geological Survey

Figure 1-5: Courtesy NASA/Jet Propulsion Laboratory

Figure 1-7: Courtesy NASA/Jet Propulsion Laboratory

Figure 2-1: Courtesy NASA/Jet Propulsion Laboratory

Figure 5-6: From Douglas, R. and F. Woodruff, 1981, in *The Sea*, v. 7, C. Emiliani, ed. New York: John Wiley & Sons. Copyright © 1981 C. Emiliani. Reprinted with permission.

Figure 5-7: Reprinted with permission from *Nature* and the authors, Barnola, J., D. Raynaud, Y. Korotkevich and C. Lorius, 1987, *Nature*, v. 329, p. 408-414. Copyright © 1987, Macmillan Magazines, Ltd.

Figure 5-9: Reprinted with permission from *Nature* and the authors, Friedli, H., H. Lötscher, H. Oescher, U. Siegenthaler and B. Stauffer, 1986, *Nature*, v. 324, p. 237-238. Copyright © 1986, Macmillan Magazines, Ltd.

Figure 6-2: Reproduced from *The Ages of Gaia, A Biography of Our Living Earth*, by James Lovelock, by permission of W.W. Norton & Co., Inc. Copyright © 1988 by the Commonwealth Fund Book Program of Sloan-Kettering Memorial Cancer Center.

Figure 7-6: From Loucks, O., The concern for acidic precipitation in the Great Lakes region, in *Acid Precipitation: Effects on Ecosystems*, F. D'Itri, ed. AnnArbor, MI: AnnArbor Science. Copyright © 1982 Butterworth-Heinemann Publishers. Reprinted by permission.

Figure S7-1: From Nuclear Energy and Fossil Fuels, by M. King Hubbert, in *1956 Drilling and Production Practice*. Copyright © 1956 by American Petroleum Institute Exploration and Production Department, Dallas, Texas. Reprinted by permission.

Figure 8-2: Courtesy US Department of Energy

Figure 8-5: Courtesy US Department of Energy

Figure 8-7: Courtesy US Department of Energy

Figure 8-8: Courtesy US Department of Energy

Figure 8-8: Courtesy US Department of Energy

Figure 8-9: Courtesy US Department of Energy

Figure 8-10: Courtesy US Department of Energy

Figure 8-11: Courtesy US Department of Energy, Yucca Mountain Project

Figure 8-11: Courtesy US Department of Energy, Yucca Mountain Project

ISBN 0-471-30629-0

Printed in the United States of America

10 9 8 7 6 5 4 3 2 1

PREFACE TO THE STUDENT

As a native Earthling, you already know a significant amount about the way the Earth works. You have seen many types of landscapes, watched seasons come and go, and perhaps witnessed exceptional events like floods, hurricanes or earthquakes. These experiences have contributed to your personal view of the Earth. The objective of this course is to introduce you to new ways of seeing and understanding your planet, through the lens of geology, which brings Eath's past and present simultaneously into focus.

The exercises in this laboratory manual do not require any previous background in geology, but you will need to observe carefully, think creatively, and write clearly. You are welcome to work with other students, but be sure not to rely on someone else to think for you.

The exercises cover a wide range of topics, from the origin of the Solar System to human consumption of Earth resources. But you will notice that a few unifying themes recur throughout:
- Earth materials are continuously recycled by processes at, above and under the planet's surface.
- These cyclical processes can be characterized in terms of *reservoirs, fluxes* and *residence times* for particular materials.
- These cycles occur over a wide range of scales in time and space, yet are linked in surprising ways.
- The complex links among these cycles are what distinguish Earth from other planets.
- Human activities have altered some of these cycles by accelerating their natural rates.
- It is possible to make inferences about the geological past because temporal (time) information can be inferred from spatial patterns.
- It is possible to make projections about the geological future based on inference about the past.

Each laboratory begins with introductory comments and a "Lab Preview" -- an ungraded self-test to help you assess how much you know about the topics to be explored. Within the exercises themselves, a running glossary in the margin provides definitions of new terms (all those <u>underlined</u> in the text). These terms are also listed in a comprehensive glossary at the end of the book (Appendix V). Each laboratory closes with suggested integrative exercises and recommended references. Useful supplementary materials included as Appendices are: Instructions for spreadsheet exercises (Appendix I); tables of metric measures and conversion factors (Appendix II); a guide to topographic and physiographic maps (Appendix III); and rock and mineral identification charts (Appendix IV).

Throughout this book, exercises of different types are identified by the following symbols:

 Study a rock, mineral or fossil specimen before answering questions

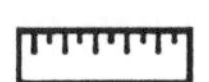 Make a measurement

 Study a map or image before answering questions

 Conduct an experiment

 Study data or graphs before answering questions

 Plot data

 Provide a written answer

 Make a sketch

 Perform simple math (you should bring a calulator with you to every class period)

 Use something you already know to make an inference

PREFACE TO THE INSTRUCTOR

A glance through this laboratory manual reveals that it departs significantly from the format and content of the traditional introductory geology lab. In particular, the conventional focus on rock and mineral identification has been replaced by a whole-Earth point of view and an emphasis on the logic and methods of the geosciences. The manual evolved from one written for use on our own campus. We had found that the "rocks in boxes" approach was not effective for most students enrolled in our general geology labs (the majority of whom were taking geology as their only science course). Students' interest in and retention of material were disappointingly low. Worse, we were ignoring global issues (climate change, plate tectonics, resource depletion) with political and economic implications -- things every Earthling should know about.

In 1992, with support from the US National Science Foundation (Division of Undergraduate Education), we instituted a radically restructured introductory geology laboratory organized around themes of Earth systems. In developing exercises for the new laboratory, we applied the following strategies :
- Pique students' interest by surrounding them with interesting visual materials;
- Allow students to frame questions and discover answers for themselves;
- Affirm that students' own experiences of Earth are valid, and use these as seeds for learning; and
- Have students collect and analyze real data and thereby become aware of limitations and uncertainties in scientific interpretations.

The result is a course defined as much by the students enrolled in it as by the instructors who lead it. Each laboratory involves a diverse range of activities, including map and image interpretation, simple experimentation, data analysis, group discussion, essay writing -- and specimen examination. Rock and mineral specimens remain an integral part of the course, but emphasis is on their interpretation and geologic context, rather than mere identification. Though time constraints require that some traditional material is left out, we believe that the new laboratory is considerably more rigorous than the old. Because students always have a clear framework for inquiry, they are able to assimilate concepts of greater complexity and subtlety.

Exercises in this manual are designed to be as "elastic" as possible, to accommodate a wide range of student and instructor backgrounds and interests. It is *not* our intent that every student at every institution complete all of the exercises. Our hope is that instructors will select those exercises most appropriate to their students and teaching style. Many of the exercises are versatile enough to be used in more than one way -- e.g. as individual written assignments, small group projects, or for class discussion, depending on the inclination of the instructor. Similarly, specimen examination/identification can assume as prominent or inconspicuous a role as the instructor considers appropriate. Although many of the exercises do build on concepts (dis)covered in preceding ones, students can begin almost anywhere in the manual if brief background information is provided by the instructor. A separate instructor's manual provides answers to exercises as well as practical and pedagogical suggestions for teaching the laboratories.

As mentioned in the Preface to the Student, the labs cover a broad spectrum of topics, from the origin of the Solar System to human use of Earth resources, but a few recurrent themes provide a unifying 'melodic line':
- Earth materials are continuously recycled by processes at, above and under the planet's surface.
- These cyclical processes can be characterized in terms of *reservoirs, fluxes* and *residence times* .
- These cycles occur over a wide range of scales in time and space, yet are linked in surprising ways.
- The complex links among these cycles are what distinguish Earth from other planets.
- Human activities have altered these cycles by accelerating their natural rates.
- It is possible to make inferences about the geological past because temporal information can be inferred from spatial patterns and process can be read from form.
- It is possible to make projections about the geological future based on inference about the past.

Recursive reference to earlier topics is a deliberate strategy throughout the book and should not be viewed as mere repetition. Students -- humans -- learn most effectively when they are able to recognize connections between new ideas and what they already know.

Each laboratory begins with introductory comments and a "Lab Preview" -- an ungraded self-test to help students assess how much they already know about the topics to be explored. Students are to return to the Preview at the close of each lab to appraise what they have learned. The Lab Preview can also be the basis of discussion and review by small groups or the class as a whole. Within the "Exercises" section of each lab, a running glossary in the margin provides definitions of new technical terms (those underlined in the text). These terms are also listed in a comprehensive glossary at the end of the book (Appendix V). Exercises of different types are identified by symbols in the margin (see Preface to the Student). The ample margins can also be used for instructor comments on students' written work. Each laboratory closes with optional integrative exercises ("Synthesis" section) and "Suggested References and Resources". In addition to the glossary, appendices include:

    I. Instructions for spreadsheet exercises (some exercises involving data analysis may be done
       with spreadsheets if computing facilities are available; this is entirely optional, however).
    II. Metric measures and conversion factors
    III. Guide to topographic and physiographic maps
    IV. Rock and mineral identification charts

Laboratory equipment and materials

This manual is designed to encourage direct observation, experimentation and analysis by students. To be most effective, the exercises should be complemented by specimens, maps and other resources -- most of which are already present in traditional geology labs.

Exercises involving specimen examination are summarized below. (Rock types in parentheses are optional).

Lab 1/ The Earth in Space
p. 10 • *Compounds condensed from the solar nebula:* Metallic iron; Olivine; Amphibole; (Meteorite)
  13 • *"Moon" rocks:* Anorthosite, basalt and constituent minerals; (Impact breccia or shatter cone)

Lab 2/ The Earth in Time
  25 • *Rocks of the Grand Canyon:* Granite; Schist; Diabase; Sandstone
  27 • *Index fossils:* Local fossils; Stromatolite; Trilobite; Graptolite; Early land plant; Ammonite
  30 • *Tree rings as time records:* Tree cores or trunk sections for ring width measurements

Lab 3/ The Earth Beneath Our Feet
  41 • *Continental v. oceanic rocks:* Granite, basalt/gabbro and constituent minerals
  49 • *Deformed rocks:* Small-scale folds; Slickensides; Cataclasite; Mylonite; Veins; Jointed rock
  50 • *Metamorphic rocks and protoliths:* Marble/limestone; Quartzite/sandstone; Slate/pelite;
      Schist/pelite; Granitic gneiss/granite; Amphibolite/basalt; Anthracite/lignite or peat
  59 • *Volcanic rocks:* Tuff; Obsidian; Pumice; Pahoehoe; Aa; Porphyry; Basalt; Andesite; Rhyolite

Lab 4/ Earth's Surface in Flux
  75 • *Clastic sedimentary rocks and structures:* Conglomerate; Sandstone; Siltstone; Shale; Diamictite;
      Ripple marks; Desiccation cracks; Cross-bedding; Graded bedding
  76 • *Local sediment samples:* e.g. River deposits; Glacial till; Loess
  78 • *Chemical sedimentary rocks:* Limestone; Evaporites; Chert; Banded iron formation
  78 • *Materials manufactured from sediments:* Glass; Brick/tile; Plaster; Concrete

Lab 5/ Earth's Aura: The Atmosphere and Hydrosphere
  91 • *Carbon-bearing rocks and minerals:* Limestone/calcite; Dolomite; Coal; Graphite; Industrial diamond

Lab 6/ The Organic Earth
 108 • *Phosphorous-bearing rocks and minerals:* Apatite; Phosphorite; Bone/bone-meal fertilizer
 113 • *Earliest lifeforms:* Stromatolites, other early fossils
 116 • *Rock types recording changes in Earth's atmosphere:* Banded iron formation; Red bed

Lab 7/ Earth Resources
 125 • *Aluminum-bearing minerals:* Feldspars; Garnet; Muscovite; Clay minerals
 126 • *Aluminum ores:* Bauxite; (Laterite)
 129 • *Metal ores:* Galena; Pyrite; Chalcopyrite; Sphalerite
 139 • *Coal:* Lignite; Bituminous coal; Anthracite; (Peat)
 140 • *Sulfur-bearing minerals:* Native sulfur; Pyrite; Gypsum/anhydrite; Low-grade coal

Lab 8/ The Imperiled Earth
 162 • *Rock types at Yucca Mtn., Nevada:* Calcite veins; Welded tuff

Other materials needed for the exercises are listed below. Most are inexpensive and easy to obtain; low-cost alternatives are suggested whenever possible. A valuable sourcebook for materials is the *Earth Science Education Resource Directory*, published by the American Geological Institute ($19.95). Write AGI Publications Center, PO Box 205, Annapolis Junction, MD 20701 or phone 301/953-1744.

MAPS/IMAGES Many maps and other figures appear in the text, but a few additional ones are useful. See also "Suggested References and Resources" section at the end of each laboratory.
- *Global views of the planets* to give students a visual overview of the Solar System (Lab 1, p. 3ff). Low-budget options: Solar System charts have appeared as supplements in *National Geographic*; check used bookstores. Planetary images are also included in most geology textbooks.
- *Shaded-relief map of the world ocean floor*, or maps of individual basins (Lab 3, p. 44ff; Lab 4, p. 74). Excellent ocean floor maps are also included periodically as *National Geographic* supplements and appear in most geology texts.
- *Global physiographic map* or world atlas (mainly Lab 3 p. 44ff, but useful throughout course).
- *Physiographic map of the US* (Lab 4, p. 69ff). We recommend the classic hand-drafted maps by Erwin Raisz. Distributed by Raisz Landform Maps, P.O. Box 773, Melrose, MA 02176. Phone: 800/242-3199; FAX: 617/662-2622.
- *Local topographic maps* (Lab 4, p. 69ff). Order from: US Geological Survey, Branch of Distribution, Box 25286, Denver Federal Center, Building 810, Denver, CO 80225. Or: Speedy Topo Service, 1705 14th St., #309, Boulder, CO 80302 (Phone 303/499-0569).

EQUIPMENT If it is not possible to obtain any of these items, the associated exercises may be omitted.
- *Hand lenses or binocular microscopes.* Useful throughout the course for specimen examination.
- *Metric rulers and drafting compasses* . For map exercises throughout the book.
- *Balances.* Not of critical importance, but useful for determining density of continental, oceanic rocks (Lab 3, p. 41); finding mass of sediment grain size fractions (Lab 4, p. 76) and aluminum cans (Lab 7, p. 127).
- *Beakers* (250 or 550 ml). For density determinations of continental, oceanic rocks (Lab 3, p. 40), a simple convection experiment (Lab 3, p. 54); and pH measurements of water samples(Lab 5, p. 92).
- *Grain size charts or sediment sieves* (Lab 4, p. 76). A supplier of inexpensive sieves is listed at end of Lab 4. If budget is very tight, simply use square pieces of wire screen of different gauges.
- *Simple permeometers* (Lab 4, p. 77). Schematics for making your own are given in Instructor Manual. Could, however, be as basic as paper cups with holes punched in bottom.
- *pH meters or litmus paper* sensitive to decimal changes in pH (Lab 5, p. 76).
- *Hot plate* for simple convection experiment (Lab 3, p. 54).
- *Miscellaneous:* Corn syrup (Lab 3, p. 54); Plexiglas sheets or overhead transparencies and impermanent pens (Lab 3, p. 56); Local water samples (rain, river, spring, well; Lab 5, p. 76).

DATA
- Local historical precipitation records for the past 80-100 yr for comparison with tree ring data (Lab 2, p.31)
- Local precipitation, evaporation and stream flow records, monthly for 1 year (Lab 5, p. 86)

We are confident that the ideas in this manual will bring new energy to your introductory laboratories. We welcome your comments and suggestions.

Acknowledgments: This project was supported by NSF grants USE-9151096 and USE-9150438. Many people have helped to see this book into print. We wish to acknowledge Joan Kalkut, Brent Peich and Bethany Brooks of John Wiley & Sons for their help with design and editing; Maryellen Cameron for early contributions to Laboratory 3; and Patrick Finkler for work on Appendix I. MGB thanks N. Grant, L. Treadwell, and J. and G. Bjornerud for their respective contributions. JMH is grateful to Susan, Gareth, and Rebecca for their continual support. ADB extends thanks to Jerry Gels, who provided data on radon emissions at the Fernald Environmental Management Project.

M. Bjørnerud, J.M. Hughes, and A.D. Baldwin, Jr.
Geology Dept., Miami University, Oxford OH 45056
August 1994

TABLE OF CONTENTS

THE EARTH
IN SPACE

How is Earth like other planets?
How is it different?

Since ancient times, humans have watched and recorded the motions of the planets. The word *planet* comes from a Greek root meaning "wanderer", because unlike the fixed stars, the planets travel the sky each night. In many cultures, these wanderers were assigned names and personalities and were thought to have control over human destinies.

In modern times, it has been possible to see the faces of the planets with telescopes and, more recently, with images transmitted by spacecraft. The planetary surface features revealed by space probes including the *Mariner, Pioneer, Viking, Voyager,* and *Magellan* series confirm the ancient belief that each of the planets has a distinct "personality".

This course is about Earth, so it may seem surprising to begin it with a discussion of the entire Solar System. But just as meeting a friend's family can help you understand your friend's traits, seeing Earth in the context of its planetary family can help us understand our planet's character.

<u>**OBJECTIVES**</u>
- To explore the large-scale structure of the Solar System
- To interpret surface features on other rocky planets
- To discover how Earth is different from other planets

<u>**OUTLINE**</u>

I. Order and structure in the Solar System
    Introduction to the planets
    Planetary spacing
    Planetary groupings
    Formation of the Solar System

II. Surfaces of the rocky planets
    Atmospheres of the rocky planets
    Mercury, the Moon and meteorites
    Landscapes of Mars and Venus

<u>LAB PREVIEW</u>   Before coming to lab,
- Answer as many of the following questions as you can
- Read Part 1: "Earth in Space" (Chaps. 1 & 2) in *The Blue Planet*, or other readings assigned by your instructor

How much do you know already about Earth's place in the Solar System?

If you can, list the planets in order of increasing distance from the Sun.

Sun

Which planets are visible to the unaided eye?

Which planet is most similar to Earth in size?

Which planet (other than Earth) bears evidence of great floods?

Why would it be impossible for a spacecraft to land on the surface of Jupiter?

What is an asteroid? A comet? A meteorite?

Why does Earth have few obvious meteorite impact craters compared with the Moon and nearby planets?

EXERCISES

I. Order and structure in the Solar System

We begin by exploring the physical characteristics of the sun and the nine planets that orbit around it. You will discover surprising patterns in the structure of the Solar System -- patterns that reveal much about the formation of the Sun and planets.

A. Introduction to the planets

1. For an overview of Earth's place in the Solar System, examine the images of the planets in your textbook and other materials provided.

2. If you weren't sure about the names and order of the planets in the Lab Preview, complete the list now.

3. In many western languages, the days of the week have names corresponding to celestial bodies: the Sun, the Moon and the 5 planets visible to the unaided eye. In English, the origins of Sunday, Monday and Saturday are obvious, but the other days of the week refer to Germanic gods, not the Roman gods for which the planets are named.

In romance languages like Spanish and French, the planetary connections are clearer. With help from people around you, complete the following table and discover how the planets still influence our daily lives.

Jove is another name for Jupiter.

TABLE 1-1.
PLANETS AND THE DAYS OF THE WEEK

<u>Day</u>	*<u>Spanish</u>*	*<u>French</u>*	Corresponding <u>body</u>
MONDAY			MOON
Tuesday			
Wednesday			
Thursday			
Friday			
SATURDAY	*(Sabado)*	*(Samedi)*	SATURN
SUNDAY	*(Domingo)*	*(Dimanche)*	SUN

scientific notation: short-hand method for writing very large or small numbers in terms of powers of 10.

logarithm: exponent to which a base number must be raised to produce a given number. For example, the logarithm of 100 in base 10 is 2.

spreadsheet: computer program for calculation, tabulation and plotting

Bode's rule: mathematical description of planetary spacing. For you to define!

B. Planetary spacing

1. Turn to Table 1-2, which summarizes the physical characteristics of the Sun and planets. Many of the entries in this table are in <u>scientific notation</u>. In this notation, numbers are written in two parts:

> a number between 1 and 10 multiplied by
> a power of 10 indicating how many digits follow the decimal point.

For example, the distance from Mercury to the Sun is 5.79×10^7 km. This is a short-hand way of writing 57,900,000 (fifty-seven million, nine hundred thousand). If the exponent, or power, of 10 were negative (5.79×10^{-7}) the number would be a very tiny 0.000000579.

Convert the following from conventional to scientific notation or vice versa:

5.3×10^9 _______________ (number of humans on Earth)

65,000,000 ______________ (number of years since dinosaur extinction)

1.6×10^{-8} _______________ (human lifetime as fraction of Earth's age)

0.0063 _______________ (fraction of Earth's water that is fresh)

2. Make a plot with the planets listed in order on the horizontal axis and the <u>logarithm</u> of their respective distances from the sun on the vertical axis. If computers are available, use a <u>spreadsheet</u> program to make your plot (see detailed instructions in Appendix I, p. 165-169). For a review of metric measurement units, see Appendix II.

3. Describe the pattern in your plot. Where (i.e. between which planets) does there seem to be a break in this pattern?

This pattern was first recognized in the late 1700s by German astronomers Johann Titius and Johann Bode, and is known as <u>Bode's rule</u>. Generalizing from the regular spacing of the known planets, they predicted that an undiscovered planet occurred at the distance corresponding to the break in the pattern. We will return to this mysterious break later.

TABLE 1-2: SUMMARY OF THE PHYSICAL CHARACTERISTICS OF THE SUN AND PLANETS

	SUN	MERCURY	VENUS	EARTH	MARS	JUPITER	SATURN	URANUS	NEPTUNE	PLUTO
Ave. distance from Sun (km)	xxx	5.79×10^7	1.08×10^8	1.50×10^8	2.28×10^8	7.78×10^8	1.43×10^9	2.87×10^9	4.5×10^9	5.9×10^9
Mass (kg)	1.99×10^{30}	3.31×10^{23}	4.87×10^{24}	5.98×10^{24}	6.42×10^{23}	1.90×10^{27}	5.69×10^{26}	8.72×10^{25}	1.03×10^{26}	1.30×10^{22}
Equatorial diameter (km)	1.39×10^6	4880	12,104	12,756	6796	142,796	120,660	51,118	49,500	2200
Density(g/cm3)	1.39	5.42	5.25	5.52	3.94	1.31	0.69	1.31	1.67	21
Length of day (in Earth days)	xxx	58.6 days	243 days	(24.62 hrs) 1 day	(9.93 hrs) 1.05 days	(10.66 hrs) 0.41 days	(17.23 hrs) 0.44 days	(16 hrs) 0.72 days	0.67 days	6.4 days
Length of year (Earth days/yrs)	xxx	88 days	224.7 days	365.26 days	687 days	11.9 yrs	29.46 yrs	84 yrs	164.8 yrs	248.5 yrs
# of moons	xxx	0	0	1	2	16	19	15	8	1
Surface character	Plasma (ionized gas)	Rocky	Rocky	Rocky	Rocky	Gaseous	Gaseous	Gaseous	Gaseous	Icy
Surface temp. (°C)	5500°	-183°(night) to +427°(day)	470°	-40° to +40°	-140° to +20	-23°	+177°	-53°	-123°	-220°
Atmos. pressure (atmospheres)	xxx	0.0 = No atmosphere	99	1	0.0052	?	?	?	?	0.0
Atmos. chemistry *Carbon Dioxide(CO_2)*	--	--	96%	0.05%	95%					--
Nitrogen (N_2)	--	--	3	78	3					--
Oxygen (O_2)	--	--	0	21	trace					--
Argon (Ar)	--	--	trace	1	1.6					--
Water (H_2O)	--	--	<1	<1	<0.1					--
Hydrogen (H)	--	--				86%	93%	84	Mainly H, He & CH_4	--
Helium (He)	--	--				14	6	14		--
Methane (CH_4)	--	--					trace	2		--

density:
mass of an object,
divided by its
volume

C. Planetary groupings

1. Look at the next three rows in Table 1-2 , which give the masses, diameters, and <u>densities</u> of the Sun and planets.

What is the total mass of the planets?

Which 2 planets account for more than 90% of this?

How does the total mass of the planets compare with the mass of the Sun?

Let this dot: • represent the size of the Earth. Draw a circle on this page illustrating the relative size of the Sun (you can draw over the text). Show your calculations. *Hint: Set up an equation of proportionality like this:*

$$\frac{\text{size of dot (in mm)}}{\text{diameter of Earth (in km)}} = \frac{\text{relative size of Sun on this page (mm)}}{\text{actual diameter of Sun (km)}}$$

2. The class will be divided into 3 groups. According to which group you are in, make a plot of the masses, diameters or densities of the planets, using the data in Table 1-2. List the planets in order on the horizontal axis and the values for mass, diameter or density on the vertical axis. (For spreadsheet instructions, see Appendix I, p. 170).

Compare your plot with those made by the other groups. On this basis, divide the planets into two broad categories. Label these on your plot.

3. In the table below, list the planets in each category and their characteristics:

TABLE 1-3.
TWO MAIN TYPES
OF PLANETS
IN THE
SOLAR SYSTEM

	<u>Category 1</u>	<u>Category 2</u>
Planets included:		
Characteristics:		
Range in masses (kg)		
Range in diameters (km)		
Range in densities (g/cm^3)		
Nature of surfaces		

4. The two types of planets are referred to as the Inner (or Rocky or Terrestrial) planets and the Outer (or Gaseous or Jovian) planets.

Which planet doesn't really fit into either category?

Where does the break between the two categories occur?

asteroids:
thousands of rocky masses orbiting mainly between the inner and outer planets, ranging in size from centimeters to about 1000 km.

5. The <u>asteroids</u> occur at the break between the two types of planets. Add the asteroids, whose average orbital distance is 4.05×10^8 km, to your original plot of planetary spacing. (For spreadsheet instructions, see Appendix I, p. 170).

What does the revised plot suggest about the nature of the asteroids?

In October 1991, the Galileo Spacecraft sent back the first close-up images of an asteroid. Its irregular shape (dimensions of about 19x12x11 km) and pitted surface suggest a violent history of collisions with other objects.

6. So why isn't there a planet at the orbital distance of the asteroids? For insight into this question, turn to Table 1-2 and find the number of moons that orbit the planets. Which two planets have the most moons? What property do these planets share? (See question I. C. 1 above)

Moons represent small pieces of planetary debris that became trapped by the gravitational pull of the relatively massive planets early in the history of the Solar System. Similarly, the asteroids are made of materials that might have formed a rocky planet. Given the position of the asteroids in the Solar System, what do you think prevented them from coalescing into a planet?

7. The distinctions between the two types of planets also provide glimpses into the early days of the Solar System. Turn again to Table 1-2 and look at the variations in surface temperatures on the different planets. What is the overall trend in temperatures as one goes from Mercury to Pluto? (We'll consider the exceptions later).

How can you explain this trend?

solar nebula:
cloud of interstellar gas thought to have collapsed under its own gravity to form the Sun and planets

element:
one of the fundamental species of matter

compound:
a chemically bound combination of elements. Water is a compound of the elements hydrogen (H) and oxygen (O).

fractionation:
separation of a material into components with distinct properties. Opposite of mixing.

D. Formation of the Solar System

1. This pattern of decreasing temperature with distance from the Sun was even more pronounced at the time the Solar System formed from a gaseous <u>nebula</u>.

The nebula consisted largely of the <u>elements</u> hydrogen (H) and helium (He) and smaller amounts of carbon (C), oxygen (O), nitrogen (N), iron (Fe), magnesium (Mg), silicon (Si), calcium (Ca), titanium (Ti), and others in trace amounts. Planets formed as the hot gaseous material cooled and condensed into solid form, much as frost on windows forms directly from water vapor in the air when temperatures are low enough.

The different <u>compounds</u> that can be made from the elements in the nebula condense at different temperatures (Fig. 1-1, next page). And because temperatures in the nebula decreased progressively with distance from the Sun (Table 1-4), the chemical composition of the condensing planets depended on that distance. In this way, the nebula <u>fractionated</u> into chemically distinct bodies.

Use Table 1-4 and Fig. 1-1 to determine the main chemical components of each planet by drawing bands whose positions and widths correspond to their temperatures of formation. This has already been done for Mercury, which formed between 1100 and 1250° C (Table 1-4) and therefore consists largely of metallic iron plus calcium-titanium and magnesium-silicon compounds. These materials have densities (symbolized by the Greek letter ρ or rho) of 7.86, 4.26 and 3.96 g/cm^3, respectively. In the third column of Table 1-4, list, in order of abundance, the main compounds that would have condensed at the temperatures of formation given for each planet. In the last column, record the densities of these compounds.

TABLE 1-4.
FORMATION OF THE PLANETS: TEMPERATURES & COMPOSITIONS

PLANET	TEMPERATURE of FORMATION (°C)	CHEMICAL COMPOSITION	DENSITY RANGE (ρ, g/cm3)[*]
MERCURY	1100 -1250	Metallic iron (Fe), CaTiO$_3$, Mg$_2$SiO$_4$	7.9 4.3, 3.2
VENUS	600-900		
EARTH	250-550		
MARS	150-250		
JUPITER	0-100		
SATURN	⁻100- 0		
URANUS & NEPTUNE	⁻150-⁻100		
PLUTO	⁻200		

In some ways Pluto is more like a comet than a planet. Also made largely of ice, comets come from the "Kuiper belt" beyond Pluto's orbit.

[*]Density values given in Figure 1-1 are for standard conditions (1 atm. pressure, 20° C), except those for H$_2$, He, & CH$_4$, whi liquid forms at boiling point, and that for H$_2$O, which is for ice at 0° C.

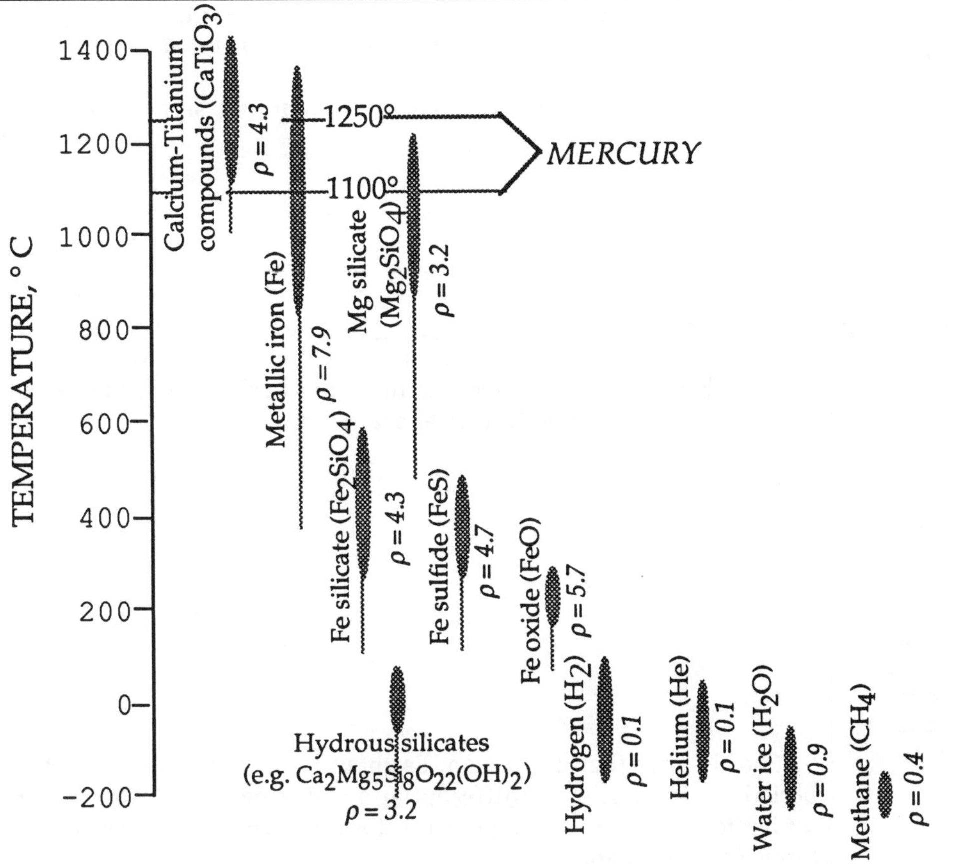

FIG. 1-1. COMPOUNDS PRESENT IN THE SOLAR NEBULA AT DECREASING TEMPERATURES
Densities (ρ) given in g/cm^3.

Sources of data: Lewis, J., 1972. *Earth and Planetary Science Letters*, v. 15, p. 286-290.
Dodd, R. 1986. *Thunderstones and Shooting Stars.* Cambridge, MA: Harvard Univ. Press, p. 161-169.

Jupiter's Red Spot is actually a long-lived cyclone-like "storm" in the gases that make up the planet's surface

Jupiter is almost massive enough to be a star, but the temperatures in its interior are not quite high enough for thermonuclear reactions to occur.

2. Explain how your entries in Table 1-4 account for the contrasting densities of the Inner and Outer planets.

On the basis of density (Table 1-2), which of the 2 types of planets would you judge to be most like the Sun in composition?

What was the probable temperature of formation of the asteroids? What does this suggest about their composition?

3. What important compound, now abundant at Earth's surface, was apparently not present when planet condensed from the solar nebula?

Speculate about how Earth might have acquired significant amounts of this compound.

4. Examine the specimens of <u>minerals</u> representing some of the solid compounds that condensed from the solar nebula to form the Inner planets. Then describe and/or sketch each specimen, taking note of characteristics like color, density, presence of visible crystals, crystal shape, etc.

mineral:
a natural crystalline solid (one in which atoms are arranged in regular three-dimensional arrays) with a well-defined chemical composition. Most rocks consist of more than one type of mineral.

5. Shortly after their formation, Earth and the other Inner planets underwent a period of <u>gravitational differentiation,</u> during which the densest materials settled to the center of the planet to form a <u>core</u> while lower-density materials formed a rocky <u>mantle</u>.

Describe an everyday example of gravitational differentiation.

gravitational differentiation
settling of materials according to density

core:
innermost and densest part of a planet. On Earth, the metallic core begins at a depth of 2900 km.

Based on the density data you entered in Table 1-4, what compounds would you expect to find in Earth's core? Earth's mantle? Indicate on the diagram below the probable compositions of the principal layers formed during early gravitational differentiation of Earth.

mantle:
layer surrounding core of the Inner planets. Earth's rocky mantle constitutes 83% of planet's volume and lies between about 40 km and 2900 km.

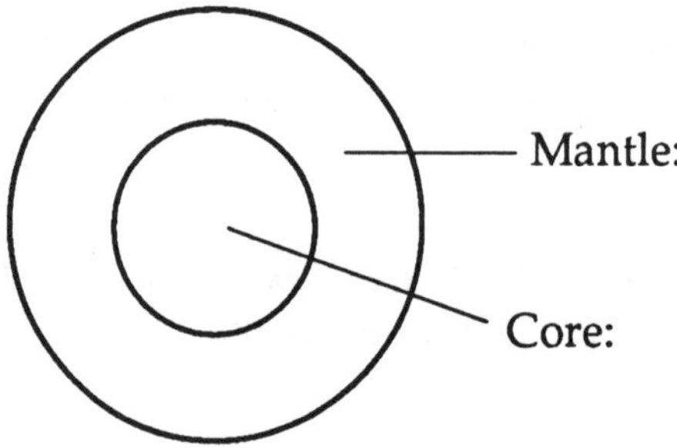

II. Surfaces of the rocky planets

Since the early 1970s, unmanned space probes from the USA and former USSR have produced startling images of the other planets and their moons. The surfaces of the rocky planets and moons bear witness to the processes which have acted on them through time. Some of the surface features resemble landscapes on Earth, suggesting that they were formed by processes like those on our planet. Other features are strange and completely unfamiliar, pointing to processes unknown on Earth.

A. Atmospheres of the rocky planets

1. Study the atmospheric compositions of the Inner planets, given in the last rows of Table 1-2. Based on the compositions of the atmospheres of Earth's nearest neighbors, what would you expect Earth's atmosphere to consist of?

How does it differ from the expected pattern?

2. Speculate about why Mercury lacks an atmosphere and Mars has only a thin one (atmospheric pressure on Mars is only 1/2% that on Earth's surface). Hint: How do Mercury and Mars differ physically from Venus and Earth?

3. Look again at the surface temperatures for the Inner planets. How does the presence or absence of an atmosphere seem to affect the constancy of surface temperatures on the Inner planets? (Which planets have the widest range in surface temperature?)

greenhouse effect: trapping of heat close to a planet's surface by gases that transmit incoming sunlight but block heat reflected back from Earth's surface. Carbon dioxide (CO_2), Methane (CH_4) and water vapor (H_2O) can all act as greenhouse gases.

Which planet seems unusually hot for its distance from the Sun?

How does the weight (pressure) of its atmosphere compare with Earth's?

The thick carbon dioxide atmosphere of Venus has caused an extreme <u>greenhouse effect</u>, making the planet oppressively hot. As on Earth, CO_2 is exhaled by volcanoes, but in the absence of plant life to withdraw it from the atmosphere through photosynthesis, the gas simply hovers over Venus as a dense cloud.

B. Mercury, the Moon and meteorites

1. Look at the images of Mercury and Earth's Moon (Figs. 1-2 and 1-3). Describe the type of surface feature that is predominant on both.

How do you think this type of feature forms?

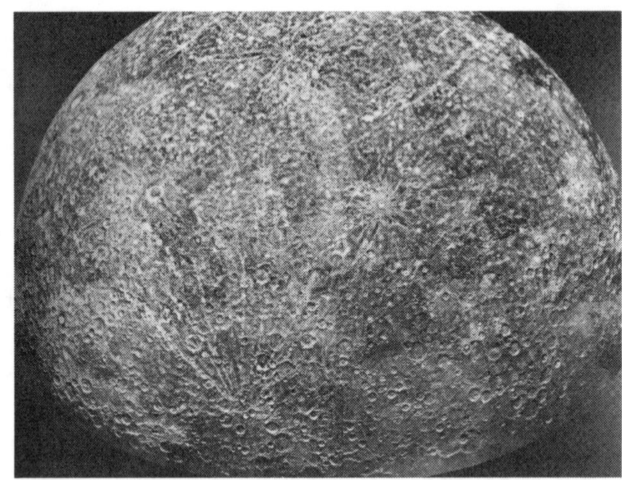

FIGURE 1-2.
PHOTOMOSAIC OF THE SURFACE OF MERCURY, FROM IMAGES OBTAINED BY *MARINER 10*
Courtesy NASA/ Jet Propulsion Laboratory

Mercury is easily visible to the unaided eye at dusk.

The two types of lunar terrain can be seen with the unaided eye on clear nights when the Moon is full.

2. Study the image of the Moon (Fig. 1-3) in more detail. Is the Moon's surface covered uniformly with the type of feature you described above?

Identify and describe two distinct types of landscape or terrain on the Moon's surface.

Which type of landscape is probably older? Explain your logic.

FIGURE 1-3.
NEAR SIDE OF THE
MOON, VIEWED
BY TELESCOPE
FROM EARTH
*Lick Observatory
photograph*

highlands (lunar):
older, heavily
cratered terrain on
the Moon

lowlands (lunar):
younger, lava-
covered terrain on
the Moon. Also
called *maria*.

rock:
aggregate of
minerals. Most
rocks consist of
more than one type
of mineral.

igneous rock:
rock formed from
the molten state
(magma)

anorthosite:
a slowly-cooled
igneous rock com-
posed mainly of the
mineral anorthite,
(Ca-rich feldspar)

basalt:
a quickly-cooled
(volcanic) igneous
rock consisting of
Ca- and Na-rich
feldspar, plus
pyroxene minerals

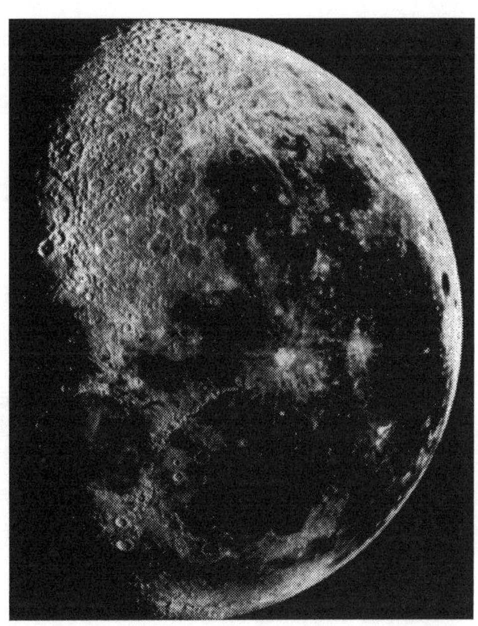

3. The two types of lunar terrain are referred to as the <u>highlands</u> and the <u>lowlands</u> or <u>maria</u> (Latin for "seas"). These terrains record distinct chapters in the Moon's history: an early phase of intense cratering, followed by a period of voluminous volcanic activity that covered up some of the early craters. Cratering continued after this, but less frequently than before.

The lunar highlands consist largely of an <u>igneous rock</u> called <u>anorthosite</u> and are thought to represent the first rocks that formed as the Moon cooled gradually from a molten sphere. Large bodies of anorthosite also occur on Earth -- for example in the Adirondack Mountains of New York. The lunar lowlands are lava plains consisting of <u>basalt</u>. On Earth, basalt is erupted by volcanoes in Hawaii, Iceland and elsewhere. It is also the predominant rock type on the seafloor.

Examine the rock and mineral specimens representative of the Moon's highlands and lowlands. (These are not actual moon rocks, but Earth rocks similar to those brought back by the *Apollo* missions). Describe the rocks, noting in particular differences in crystal size. Which rock type has larger crystals? What does crystal size reflect about the cooling history of an igneous rock? Explain.

See also Appendix IV: Rock and mineral identification charts.

ejecta:
rock material thrown out explosively from the site of an impact

Most meteorites that fall to Earth are probably asteroids with Earth-crossing orbits. A few are thought to be ejecta from impacts on Mars and the Moon.

Because they are remnants virtually unaltered since the time that the planets formed, meteorite fragments like those found at Meteor Crater are important sources of information about the earliest days of the Solar System.

iridium:
element 77, very rare on Earth, but found in higher concentrations in meteorites

4. If the surfaces of Mercury and the Moon, both near-neighbors of Earth, are substantially covered by impact craters, why do you think such craters are not conspicuous on this planet? Propose at least three possible explanations.

5. Although impact craters are not as common on Earth as on other rocky planets and moons, they do occur from place to place. More than 100 terrestrial impact sites have been identified, including the well-known Meteor Crater in Arizona.

Examine the aerial photograph of Meteor Crater (Fig. 1-4). The crater is relatively recent -- less than 50,000 years old. Compare the crater rim and <u>ejecta</u> deposits with those around craters on Mercury and the Moon (Figs. 1-2, 1-3). Most of these craters are more than 3 billion years old. Why do you think the rim of Meteor Crater is smooth and subdued compared with those of the far older craters on Mercury and the Moon?

How does this help to explain the relative rarity of impact craters on Earth?

6. Although the processes that act continuously on Earth's surface may alter, erase or bury impact craters, impact sites can be identified even where there is no obvious crater. Meteorites traveling at supersonic speeds permanently modify the rocks they strike, creating <u>impact breccias</u> (chaotically fragmented rock) and <u>impact melts</u> (glass formed at the moment of impact).

Meteorites also leave behind distinctive chemical traces, including relatively high concentrations of <u>iridium,</u> an element that is extremely rare on Earth. In 1980, the detection of high iridium concentrations in rocks at the boundary between the Cretaceous and Tertiary periods -- the geologic moment when the dinosaurs went extinct -- led to the proposal that a huge meteorite struck Earth 65 million yrs ago.   It took ten years, however, before a trail of impact breccias and melts lead to a crater of the right age. A buried crater at Chicxulub, on the northern edge of Mexico's Yucatan peninsula, is now thought to have been the site of the cataclysm. At the time of the impact, the area was beneath a shallow sea.

What special consequences would follow an impact in the oceans? What evidence for such an event might be preserved in the rock record?

FIGURE 1-4.
AERIAL VIEW OF
METEOR CRATER,
NORTHERN
ARIZONA
Crater is about
1.5 km in diameter.
Courtesy D. Roddy,
US Geological Survey

shield volcano:
broad, gently-
sloping volcano
built from many
thin layers of
basaltic lava

In the late 1800s, an
Italian astronomer
announced that he had
seen canals on the
surface of Mars. The
lines he thought to be
canals may have
included Valles
Marineris, *a canyon so*
long it would stretch
from New York to LA.

Many features indicate
that Mars once had
running water -- and
even catastrophic floods.
Now its water is locked
in ice, mostly in polar
ice caps but also as
permafrost in the thin
Martian soil.

C. Landscapes of Mars and Venus

1. Although Mars has more impact craters than Earth, it has fewer than Mercury or the Moon. Its surface features are varied, reflecting the wide range of processes active or once active on the planet.

The most prominent topographic features on Mars are giant <u>shield volcanoes</u> similar in some respects to the Hawaiian volcanoes on Earth. But Martian volcanoes are much larger than their Earthly counterparts. Olympus Mons -- the largest volcano not only on Mars but in the entire Solar System -- is 27 km high and 600 km wide at its base. Hawaii's Mauna Loa, Earth's largest volcano, is only 9 km high and 225 km at its base (measured from the sea floor) .

Make a sketch showing the relative sizes of Olympus Mons and Mauna Loa. Indicate the scale of your drawing.

2. Study the image of the surface of Mars (Fig. 1-5). Describe the features on the left side of the image. Do they remind you of features you have seen on Earth? How do you think they might have formed?

Is this consistent with the present range of surface temperatures on Mars (Table 1-2)? Explain.

3. The high surface temperatures and dense atmosphere of Venus have hindered exploration of the planet's surface. The only successful landing on Venus was made in the mid-1970's by the unmanned Soviet spacecraft *Venera*, which managed to transmit several images back to Earth before being crushed by the pressure of Venus' atmosphere. Little was known about the surface of Earth's sister planet until 1978, when the *Pioneer Venus* spacecraft sent back cloud-penetrating radar images of its face.

Pioneer revealed distinct high-standing areas (*terrae*, Latin for "land") on Venus' surface. These initially appear to be analogous to Earth's continents, but a <u>histogram</u> showing the percent of the surfaces of Venus and Earth that lie at various elevations (Fig. 1-6) reveals that the similarity is only superficial.

Describe the patterns shown in the histograms. How is Earth's topography fundamentally different from Venus'?

histogram:
plot showing the frequency with which particular values of a variable occur

FIGURE 1-6.
DISTRIBUTION OF SURFACE ELEVATIONS ON EARTH AND VENUS

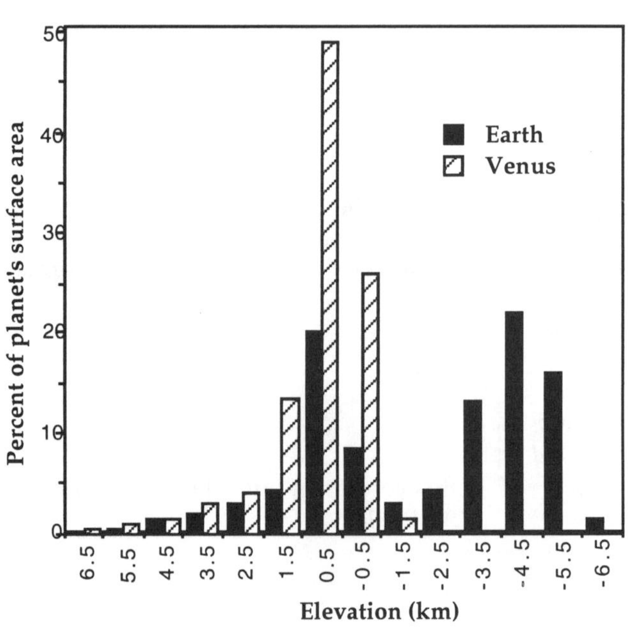

The surface of Venus is now known in more detail than the topography of Earth's ocean floor!

Venus is named for the Roman goddess of beauty, and many regions and features on the planet are named after real and mythical women. For example, there is a crater named Cleopatra *and one named* Sappho, *after the Greek poet.*

In some regions of Venus, lava has cut meandering channels that resemble river channels on Earth.

4. In 1990, the *Magellan* spacecraft began transmitting much higher-resolution radar images of Venus back to Earth. These images portray objects as small as 120 m (400 ft) and reveal an astoundingly varied planet with features both familiar and unfamiliar to Earthlings.

Study Figure 1-7, a *Magellan* image of lava plains in the Lada region (47°S, 25°E). Describe the form of the lava flows in the image. What does the character of the flows suggest about their behavior?

How might climatic conditions on Venus affect the behavior of lava at its surface? (Review Table 1-2)

Compared with the surfaces of Mercury (Fig. 1-2) and the Moon (Fig. 1-3), does the landscape in this region of Venus appear to be relatively young or old? Explain.

FIGURE 1-7.
MAGELLAN RADAR IMAGE OF LAVA PLAINS IN THE LADA REGION OF VENUS
Lava is flowing from top to bottom, through a low ridge running from left to right. Width of scene is about 600 km
Courtesy NASA/JPL

III. Synthesis

A. Preview review
Return to the Lab Preview and complete any questions you were unable to answer before.

B. Young and old planetary surfaces
All of the planets formed at about the same time, but since their formation they have had very different histories. The surfaces of some planets are frozen records of their earliest days, while the surfaces of others have been renewed many times by volcanic, tectonic and climatic processes.

Based on what you have learned in this laboratory, list the Inner planets and Earth's Moon in order of the relative ages of their surfaces. Briefly explain your logic.

Oldest surface . . . . Youngest surface

C. Earth's uniqueness
In some respects, Earth is just another planet in the Solar System. In other respects, it is completely different. As a way to summarize the main ideas addressed in this laboratory and to anticipate the ones to follow, identify characteristics that Earth shares with one or more planets and the important ways in which it is unlike all other planets.

Characteristics shared
<u>with other planets:</u>

Characteristics
<u>unique to Earth:</u>

SUGGESTED REFERENCES AND RESOURCES

Solar system overview:

 Aaron, G., ed., 1990. *The Voyager Poster Book*. Philadelphia: Running Press. 20 pp.
Seventeen color posters of the planets (except Pluto) and their moons. Back of each poster has synoptic information about the featured body. Running Press, 125 S. 22nd St., Philadelphia, PA, 19103. Telephone: 800/345-5359.

 Malin, Stuart, 1989. *The Greenwich Guide to the Planets*. New York: Cambridge University Press. 96 pp.
A concise and readable book that integrates current views on the Solar System with the history of its exploration since ancient times.

 Smith, B. A., 1990 (August). Voyage of the Century, *National Geographic*, v. 178, no. 2, p. 49-65.
A good summary of the Voyager missions, with stunning images of the Jovian planets and their moons, accompanied by a color poster with illustrations of and information about the Sun, planets, and asteroids. A separate article on Neptune appears in the same issue.

 *ORBITS: Voyage through the Solar System*
An interactive hypermedia atlas of the Solar System for PCs. Mouse driven and very easy to use. Priced at about $50. Software Marketing Corporation, 9831 South 51st St., Bldg C-113, Phoenix, AZ 85044. Telephone: 602/893-2400.

Surfaces of the rocky planets:

 Greely, Ronald, 1994. *Planetary Surfaces* (2nd edition), New York: Chapman & Hall, 288 pp.
A comprehensive and newly updated guide to the interpretation of surface features on the rocky planets and moons.

 NASA/Jet Propulsion Laboratory, 1993. *Magellan: Revealing the Face of Venus*. 25pp. ISBN 0-16-042197-7.
A beautifully illustrated, lucidly written description of the *Magellan* mission and its findings. For sale by US Government Printing Office, Superintendent of Documents, Mail Stop SSOP, Washington, DC 20402-9328.

 Newcott, W., 1993 (February). Venus Revealed, *National Geographic*, v. 183, no. 2, p. 36-65.
A good summary of the *Magellan* mission, with characteristically lavish illustrations.

THE EARTH IN TIME

*How can we reconstruct
events in the earth's past?*

How do we know anything about the past? You carry information about the past in your memory, but that information includes only your own experiences. Written documents -- from newspapers to cave-drawings -- preserve information about the past and give us glimpses into times and places we did not experience ourselves. Still, these documents record only events that were observed by humans.

Making inferences about events that occurred without human witnesses is the central theme of Geology. Earth's landscapes, rocks, waters and lifeforms all reflect events that occurred at times in the past. They are "documents" containing encoded information about earlier times. Decoding these documents is the key to reconstructing the past. As a starting point, let us assume that the same natural processes and physical laws that shape Earth's surface now also operated in the past. This simple idea is called <u>uniformitarianism</u> and can be summarized by the phrase "the present is the key to the past". While the concept may seem obvious today, it was revolutionary when first introduced in the late 1700's. Until then, the prevailing view was that the form of Earth's surface reflected cataclysmic processes that had ceased to occur.

A uniformitarian point of view is the first step in reading Earth history from natural features. With an understanding of the modern Earth, it is possible to make inferences about the ancient Earth.

<u>OBJECTIVES</u>
- To learn to read time information from rocks and landscapes
- To understand how absolute ages can be assigned to events in Earth's distant past

<u>OUTLINE</u>

I. Determining relative ages of events
  Rules for inferring relative timing
  Applying the rules of relative timing

II. Correlation of fragmentary records
  Unconformities
  Index fossils
  The fossil-based geologic time scale

III. Determining absolute ages of events
  Tree-ring analysis
  Radioactive isotopes as clocks

IV. The geologic time scale
  Integrating relative, absolute ages
  The age of the Earth

<u>LAB PREVIEW</u>   Before coming to lab,
- Answer as many of the following questions as you can
- Read the following sections from *The Blue Planet* :
    Introduction: "Uniformitarianism"
    Chapter 7 (Earth's Evolving Crust):
        "Sedimentary rocks"
        "The Geologic Column"
        "A closer look: Radioactivity & the measurement of time"
    Or other readings assigned by your instructor

How much do you know already about the history of the Earth?

What do the rocks and landscapes in your area reveal about events in the recent and distant past?

How can these events be placed sequentially within a global time scale?

How old are the oldest rocks found on Earth's surface? Have any rocks survived from the time Earth formed?

What is the age of the Earth? How has this estimate been made?

EXERCISES

principle of superposition: principle that in a layered sequence of rocks that accumulated at Earth's surface, the oldest rocks are at the bottom

sedimentary rock: rock composed of recycled pieces of earlier rocks, deposited by water, wind or ice

principle of cross-cutting relationships: principle that if feature A is cut or modified by feature B, then A must be older

I. Determining relative ages of events

Using simple logic, you can develop general rules for reconstructing sequences of events in the geologic past.

A. Rules for inferring relative timing

1. a) Imagine that you have been on a long trip and that each day while you were gone your roommate put your mail in a growing pile on your desk. Assuming that the pile has not toppled over or been otherwise disturbed, where would you find the oldest mail? The most recent mail?

This is a simple illustration of one of the most fundamental rules for reconstructing geologic events: the <u>principle of superposition.</u> Note that the principle applies only to geologic materials that were laid down sequentially in horizontal layers at Earth's surface -- i.e. to <u>sedimentary rocks</u> and to volcanic igneous rocks.

b) Draw a simple sketch illustrating how the principle of superposition can be used to interpret the geologic history of your area or of the specimen provided.

2. a) You are now reading through the towering stack of mail and accidentally knock it over, destroying the original order of accumulation. An undated letter from a friend has an illegible postmark, and you wonder when it was written. You see that the letter's postmark is smeared because someone spilled coffee on it. Whom do you blame: the friend who wrote the letter or your roommate? Why?

b) This illustrates a second fundamental rule for reconstructing geologic events: the <u>principle of cross-cutting relationships.</u> Draw a sketch illustrating how the principle of cross-cutting relationships can be used to interpret the geologic history of your area or of the specimen provided.

principle of inclusion:
A rock that contains (includes) pieces of other rocks is younger than those pieces. (Conversely, the pieces are older than the rock that contains them)

In Lab 1, you also used the principle of cross-cutting relationships to infer the relative ages of the heavily cratered lunar high-lands and the lava-covered lowlands.

3. a) The letter includes a Polaroid photograph taken at an event that you know occurred 2 weeks ago. What constraints does this place on the possible times the letter could have been written?

b) This illustrates a third fundamental rule for reconstructing geologic events: the <u>principle of inclusion</u>. Draw a sketch illustrating how the principle of inclusion can be used to interpret the geologic history of your area or of the specimen provided.

B. Applying the rules of relative timing
1. The principle of cross-cutting relationships has been the basis for reconstructing the histories of the rocky planets and moons. Examine the image of the Alpha Regio area of Venus (Fig. 2-1), showing several dome-shaped hills thought to represent mounds of very viscous (sticky) lava. (Similar volcanic domes occur on Earth; see Lab 3). Use the principle of cross-cutting relationships to determine the relative ages of: Domes A, B and C, crater D, and Fracture E. Explain your logic and note all cases in which the ages of any two features cannot be resolved.

FIGURE 2-1.
MAGELLAN RADAR IMAGE OF VOLCANIC DOMES IN ALPHA REGIO, VENUS
Domes are approximately 25 km in diameter.
Courtesy NASA/JPL

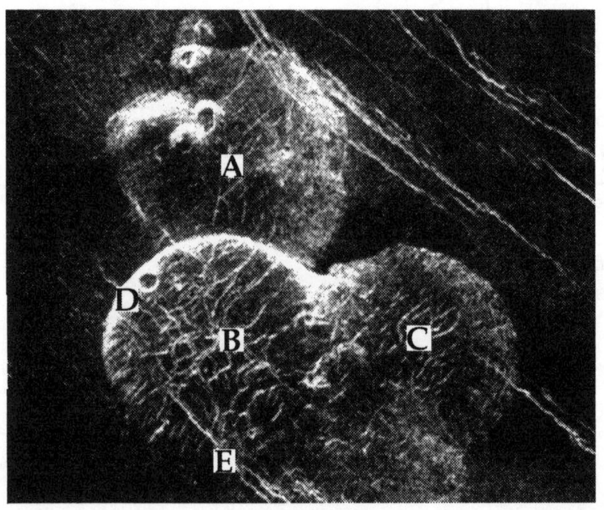

cross section:
schematic drawing
of a vertical cut
into the subsurface

granite:
igneous rock
intruded into
subsurface; main
minerals: sodium/
potassium feldspar,
quartz and mica

protolith:
original rock from
which a metamorphic
rock formed

metamorphic rock:
rock modified by
high temperatures
and/or pressures

schist:
metamorphic rock
formed from fine-
grained sedimentary
rock (mudstone)

FIGURE 2-2.
GEOLOGIC CROSS-
SECTION OF THE
GRAND CANYON

diabase:
intrusive igneous
rock with composi-
tion similar to
that of basalt

sill:
igneous intrusion
emplaced parallel
to rock layering

fault:
rock fracture along
which slip has
occurred

2. Rocks exposed in the Grand Canyon in Arizona record many episodes of modification and renewal of Earth's surface. At least 6 distinct geologic events are recorded in the cross section of the Canyon below (Fig. 2-2). These include:

- Deposition of sedimentary sequence A
- Intrusion of granite B
- Deposition of sedimentary protolith of schist C
- Intrusion of diabase sill D
- Deposition of sedimentary sequence E
- Displacement on fault F

Determine the order in which these events occurred and list them below, beginning with the earliest. Explain which timing rule(s) you used to constrain the ages and note any cases in which the relative ages cannot be resolved.

Event	Timing rule applied to constrain age

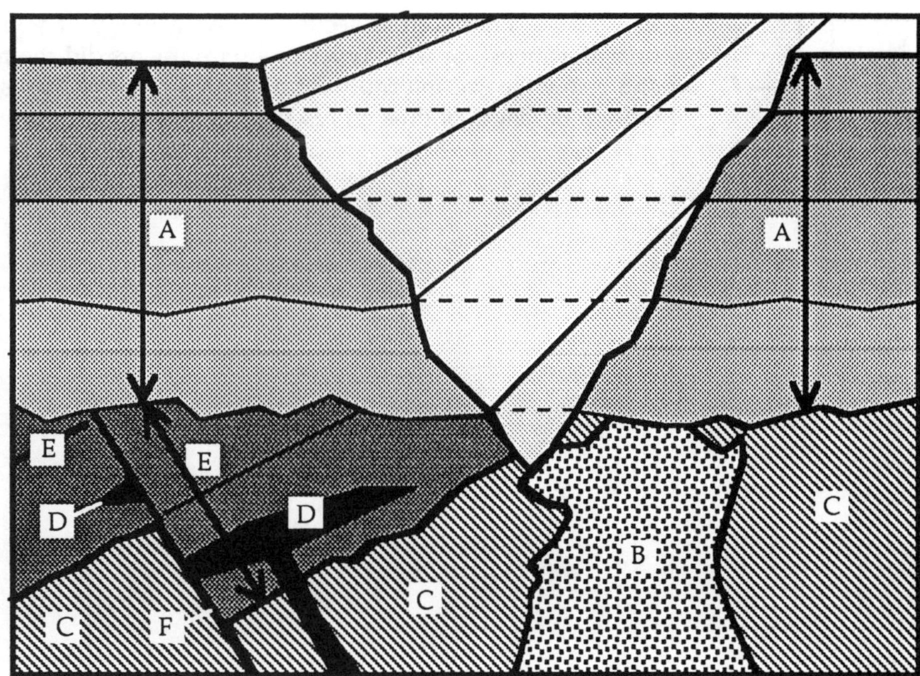

3. Examine the specimens representing rock types exposed in the Grand Canyon.

II. Correlation of fragmentary records

You have now developed and used criteria for reconstructing the sequence, or chronology, of geologic events in a given area. Though you could work out such event chronologies in many different places, it would be difficult to know how these local chronologies were related to each other. Understanding Earth history on a global scale requires <u>correlation</u> of events from place to place.

A. Unconformities

One difficulty in correlation is that no single place on Earth -- not even the Grand Canyon -- has a complete, uninterrupted record of all of geologic time. Instead, there are gaps representing periods when erosion, or at least no deposition, occurred. Such a gap in the sedimentary record is called an <u>unconformity</u> and can be likened to pages missing from an historical document.

1. Identify and label at least one major unconformity depicted in the cross section of the Grand Canyon (Fig. 2-2). Explain why this surface must represent a long period of erosion. (What events must have taken place in the time between the formation of the rocks below the unconformity and those above it?)

2. Draw a simple cross section of the geology of your area and identify any unconformities. In what types of environments did the rocks or sediments above and below the unconformity form? Can you make any estimates of the amount of erosion or length of time represented by the unconformity(ies)? Explain.

fossil:
preserved remains
or traces of a
living organism

index fossil:
fossil of an
organism whose
species existed for
a geologically
brief time period.
Index fossils are
very useful in
correlation.

B. Index fossils

Though the record at any one place is incomplete, <u>fossils</u> make it possible to recognize overlaps in the fragmentary records at different places. By the mid-1800s, the study of fossils (paleontology) had revealed that there is a globally consistent order in the appearance and disappearance of different lifeforms in the sedimentary record. With fossils acting as global "page numbers", the fragmentary geologic records from different places can be ordered and divided into time intervals recognizable worldwide.

The most useful fossils for correlation are those of plant and animal species that lived for only a geologically brief period of time. Such fossils, called <u>index fossils,</u> are thus diagnostic of a particular time in the geologic past. Species that changed little through time do not make useful index fossils.

1. Consider a collection of cast-off objects from this century and imagine them as fossils. List 2 items that would make good "index fossils" for a particular decade in the 1900's. Conversely, list 2 items that were available during that decade but would *not* make good index fossils. Explain.

Decade:

<u>Good index fossils</u>   <u>Poor index fossils</u>

2. Examine the index fossils on display in the laboratory. Select one and make a sketch of it. What sort of organism was it? In what type of environment did it live? How was it preserved?

C. Fossil-based geologic time scale

1. By the late 1800s, careful correlation of fossils in sedimentary sequences from around the world had made it possible to create a global geologic time scale. This exercise is a simple illustration of how this was accomplished.

Columns A-D in the figure below depict fossil-bearing sedimentary sequences at four sites. The rock records at each site are continuous except at the zig-zag lines, which represent unconformities. All of the fossils can be considered index fossils. By comparing the sequences, determine the relative ages of the index fossils and enter them in the proper 'global' order in the time-scale at right (1 = oldest; 6= youngest).

FIGURE 2-3.
ILLUSTRATION OF USE OF INDEX FOSSILS FOR GEOLOGIC CORRELATION

ammonites:
related to modern squid, the detailed shell ornamentation of these animals makes them superb index fossils for the Jurassic and Cretaceous periods (see Table 2-1)

graptolites:
possible predecessors of earliest vertebrates; excellent Silurian index fossils

stromatolites:
fossil algal mats, first appeared in the mid-Precambrian

trilobites:
early arthropods (like modern crustaceans). Cambrian, Ordovician index fossils

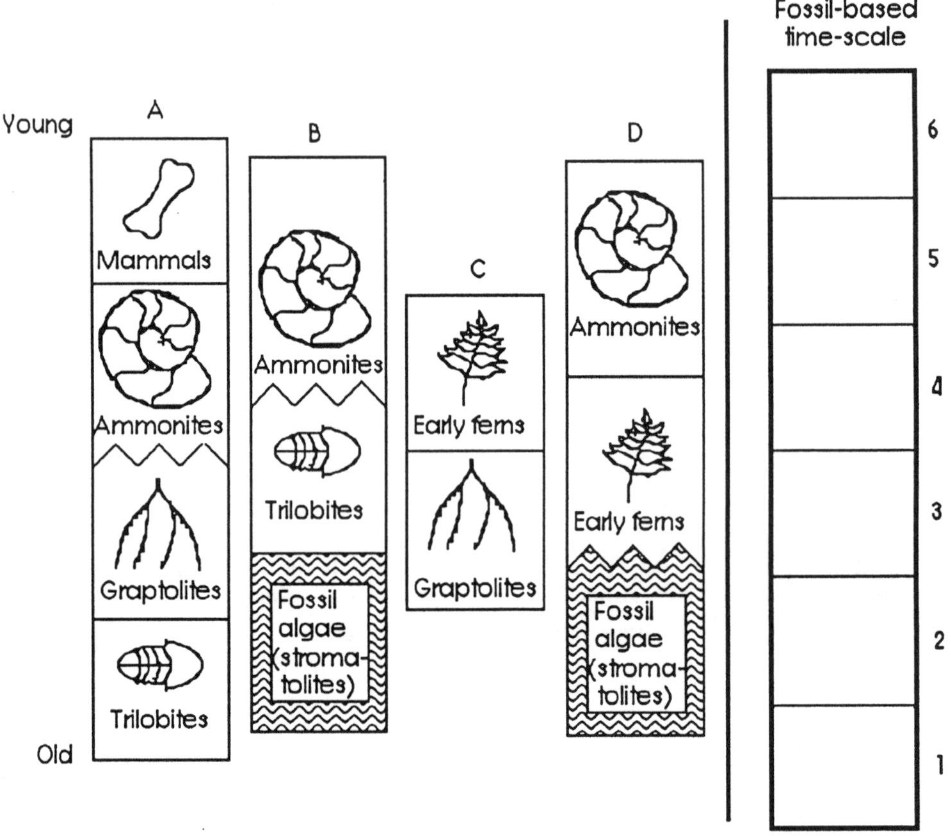

mass extinction: abrupt disappearance of many species from the fossil record

2. The main divisions of the modern geologic time scale (Table 2-1) were based on the known succession of fossil life forms and had been established by the end of the last century. Many of the divisions were defined on the basis of <u>mass extinctions</u> in the fossil record. The end of the Mesozoic Era, for example, was defined as the time of simultaneous extinction of the dinosaurs and many other species.

To which divisions of geologic time do the rocks and deposits of your area belong? Circle or highlight them on the table below, then label the appropriate ages on rock units in the cross section you drew for exercise II.A.2.

TABLE 2-1. THE FOSSIL-BASED GEOLOGIC TIME SCALE

** In North America, the Carboniferous is subdivided into the Mississippian and Pennsylvanian periods.*

EON	ERA	PERIOD	EPOCH	Notable lifeforms & events
Phanero-zoic	Cenozoic	Quaternary	Holocene Pleistocene	*Written history* *Ice age*
		Tertiary		*Mammals proliferate*
	Mesozoic	Cretaceous		*Extinction of dinosaurs*
		Jurassic		*First birds*
		Triassic		*Reptiles proliferate*
	Paleozoic	Permian		*Mass extinction: 95% of existing marine species disappear*
		Carboniferous Pennsylvanian* Mississippian*		*Much plant matter preserved in widespread coal swamps*
		Devonian		*Land plants begin to flourish* *Fish appear*
		Silurian		*First coral reefs*
		Ordovician		*Proliferation of marine life*
		Cambrian		*Sudden appearance of diverse marine life with shells & skeleta*
Pre-cambrian	Proterozoic			*Few records of life other than fossil algal colonies*
	Archean			

*Annual rings are
preserved so well in
some petrified wood
that inferences about
tree growth patterns
can be made. A recent
study of 20 million
year old fossil wood
from Japan allowed
reconstruction of the
frequency of ancient
volcanic events.*

*Tree-ring analysis
does not usually
involve cutting down
trees; instead, a
special auger is used
to extract a slender
core for study -- the
tree is unharmed.*

III. Determining absolute ages of events

Fossils make it possible to establish worldwide geologic time divisions, but indicate nothing about how long ago these time periods occurred or how long they lasted. The key to determining the <u>absolute ages</u> of geologic events -- and of Earth itself -- is to find processes that occur at known rates and leave behind measurable products that accumulate through time. By measuring the amount of the product, it is possible to estimate the time elapsed since the process began.

A. Dendrochronology/Tree-ring analysis *(Greek: dendro = tree; kronos = time)*
Tree growth is a familiar process that leaves behind measurable records of time. Tree rings represent new material added to the tree during the growing season each year. In this exercise you will explore how tree rings are "archives" of climatic information.

1. Study one of the tree cores or sections provided in the lab. The inner rings are the oldest. Each year's growth includes one light and one dark band, representing early and late season growth, respectively. Examine the rings with a hand lens, then describe or sketch the differing characteristics of the light and dark wood.

2. About how old was this tree when it was cut down?

3. Compare your age estimate with those made by other students (all sections came from the same tree). Discuss sources of uncertainty in your age determination. What problems did you encounter in estimating the tree's age? What assumptions did you make? How might you improve the accuracy of your age determination?

4. Measure the width of each of the rings (in millimeters) and record the values for each year before present. It might be easiest to lay the edge of a piece of paper along a radius of the trunk section, then mark the ring widths on the paper. Record your data in the table on the next page.

TABLE 2-2:
TREE RING DATA

Like tree rings, ice caps preserve annual time records. Alternating layers of ice and dirt represent winter snows and summer melting. Ice at the bottom of cores taken from Greenland and Antarctica fell as snow more than 200,000 yrs ago! (see Lab 5).

Ring #	Year*	Width (mm)		Ring #	Year*	Width (mm)

**Based on year tree was cut down*

Like people, trees grow fastest when they are young.

Rings from living trees provide climatic records up to 6000 years old. Correlation of rings in living trees with rings in logs preserved in bogs and prehistoric buildings has extended the tree-ring record to more than 13,000 years.

5. In which years did the tree apparently grow fastest? What factors might determine the rate of tree growth?

6. Make a plot with ring width on the vertical axis and years AD on the horizontal axis, and compare it with historical precipitation records from your area, if these are available. (See Appendix I, p. 171 for spreadsheet instructions).

7. Discuss the correspondence between the precipitation and ring width curves. Do years of heavy precipitation coincide with years of rapid tree growth? Do years of light precipitation coincide with years of slower tree growth? Does precipitation seem to be the main factor affecting ring width? If not, what other factors may have been more important? How could you test their importance?

**radioactive
decay** or
radioactivity:
spontaneous break-
down of unstable
isotopes into heat,
subatomic particles
and other isotopes

isotope:
atom of an element
that can have
varying numbers of
neutrons in its
nucleus

*Addition of annual
rings in a tree is a
linear process -- that
is, a constant number
of rings is added in a
given time interval.*

*Radioactive decay,
in contrast, is an
exponential process
-- the amount of
material decaying in
a given time inter-
val is not constant
but depends on the
amount present.*

B. Radioactive isotopes as natural clocks

Unfortunately, direct time-records like tree rings and glacial layering are useful only for the relatively recent past. How can we determine the ages of older events? Like the growth of trees or the accumulation of snow on glaciers, radioactive decay is a natural process that occurs at a known rate, creates measurable products, and can be used to date geologic materials and events. To understand natural radioactivity, we need a quick review of chemistry.

All matter consists of elements or combinations of elements. The smallest subdivision of an element that still has properties of that element is an atom. Atoms consist of a central nucleus, containing protons and neutrons, and an outer cloud of orbiting electrons. The number of protons in the nucleus of the atom -- the atomic number -- is the characteristic that gives an element its identity. For example, all oxygen atoms have 8 protons; all gold atoms have 79 protons. In contrast, the number of neutrons in atoms of some elements can vary, and the various forms of such elements are called isotopes. Some isotopes are stable and remain unchanged through time. Others are unstable and tend to break down, or decay over time, by giving off subatomic particles and energy. Some elements that can occur as unstable isotopes are carbon (C), potassium (K) and uranium (U).

So how can unstable, or radioactive, isotopes be used to determine the ages of rocks? In the same way that the number of annual rings gives the age of a tree, the key is to know the rate at which a process occurs, measure its resultant product and calculate how long the process was going on. However, the process of radioactive decay is rather different from the yearly growth of trees. Atoms of unstable isotopes decay in numbers that are a constant proportion of the unstable atoms present, so the number of atoms that decay in a particular time interval is proportional to the number remaining.

Imagine that a parent decides to give his daughter exactly half of his money each day. The first day he gives her half of his total wealth. The next day he gives her half of what remains, or one-fourth of the original. The following day he gives her half of the new remainder -- one-eighth of the original -- and so on. Assuming that the father could continuously subdivide what remained of his wealth, he would never run out completely, though his reserve would grow very small. Conversely, the amount of money his daughter held would always increase, though the amount she received each day would grow smaller and smaller. At any time, it would be possible to determine how many days the parent had been giving money to his daughter by comparing the amounts held by each.

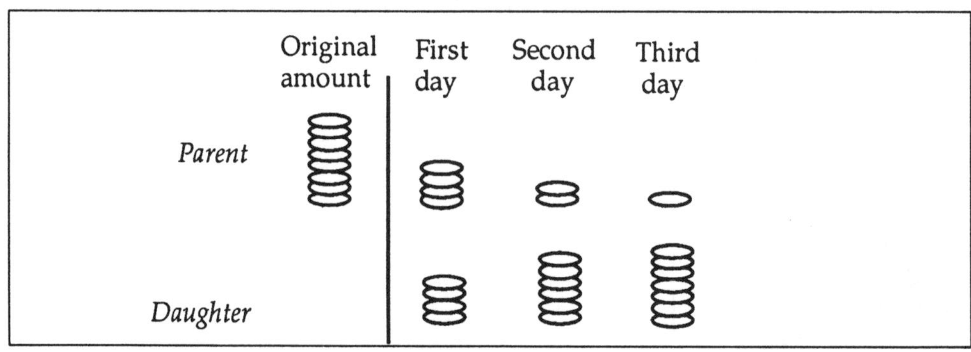

ratio:
the quotient of two numbers

1. What is the <u>ratio</u> of the amount of money held by the daughter to the amount held by the parent:

<u>Daughter's $/Parent's $</u>

At the beginning:

At the end of the 1st day:

At the end of the 2nd day:

At the end of the 3rd day:

2. If the parent started with twice as much money (16 coins instead of 8), what would be the ratios of the daughter's to the parent's money each day? Find out by completing the following table.

	Amount held by parent	Amount held by daughter	Daughter's $/ Parent's $
At the beginning:			
At the end of the 1st day:			
At the end of the 2nd day:			
At the end of the 3rd day:			

3. Would the ratio of the amounts held by the daughter and parent on any given day be different if the parent originally held four times as much money? Ten times as much money? Explain.

half life:
time required for half the amount of an unstable parent isotope to decay to a daughter isotope

atomic mass number:
sum of the number of protons and neutrons in the nucleus of an atom.

The analogy of the parent and daughter illustrates the essential principles behind isotopic dating. The parent gave his daughter half his money each day; similarly, every radioactive isotope has a characteristic <u>half-life</u> -- the length of time for half of the original or <u>parent</u> atoms to decay to <u>daughter</u> atoms. The half-lives of some geologically useful isotopes are listed in Table 2-2. The number after each element name is the <u>atomic mass number</u> of the isotope.

As you showed above, the ratio of the amount of accumulated daughter isotopes to the remaining parent isotopes does not depend on the original amount of parent material. So comparison of the relative amounts of daughter and parent isotopes allows determination of time elapsed since the parent isotope began to decay. More detailed discussion of the uranium-lead decay series can be found in Ch. 8.

TABLE 2-3.
GEOLOGICALLY USEFUL ISOTOPE SYSTEMS

Parent isotope	Stable daughter isotope	Half-life	Dateable materials
Carbon-14	Nitrogen-14	5730 yrs	Organic matter <50,000 yrs old
Potassium-40	Argon-40	1.5×10^9 yrs	Igneous, metamorphic rocks
Uranium-235	Lead-207	7.13×10^8 yrs	Igneous, metamorphic rocks
Uranium-238	Lead-206	4.5×10^9 yrs	Igneous, metamorphic rocks

Most sedimentary rocks cannot be dated directly by isotopic methods, but their ages can be constrained by isotopic ages of cross-cutting igneous rocks.

Unlike other unstable isotopes, ^{14}C is continuously replenished (by activity in the upper atmosphere).

Wood, bone, cloth, and even water (which contains dissolved CO_2) can be dated by the ^{14}C method. Deep ocean water gives ^{14}C ages of >3000 years -- the time since it last circulated through the atmosphere.

^{14}C dating of the Shroud of Turin by three independent laboratories revealed that it is a medieval relic.

In the analogy of the parent and daughter, the "decay" process began when the parent began to give his daughter money. For most geologically useful isotopes, like the potassium-argon (K-Ar) and uranium-lead (U-Pb) systems, the starting point is the time at which igneous or metamorphic rocks cooled to a particular temperature. For carbon-14 (^{14}C), the isotopic "clock" is begun when an organism dies and ceases to take atmospheric carbon into its tissues.

The length of the half-life determines the time-span over which a particular isotope system is useful for geological dating purposes. For example, the ^{14}C method, with a half life of only 5730 years, can only be used to date materials less than about 50,000 years old because after 9-10 half-lives have elapsed the amount of parent material remaining is immeasurably small. Conversely, isotopic systems with long half lives are not useful for very young rocks since the amount of accumulated daughter material would be too small to measure.

4. What was the "half-life" in the parent-daughter analogy? Explain.

5. Uranium-lead analysis of an igneous rock shows that the ratio of daughter ^{207}Pb to parent ^{235}U isotopes is 7. How many half-lives have elapsed since the rock cooled and crystallized? Explain your answer.

In another igneous rock, the $^{207}Pb/^{235}U$ ratio is 15. How many half-lives have elapsed since it crystallized? How many years does this represent? Explain. Hint: Extend the table you began in question III. B. 2 on the previous page. You could also use a spreadsheet to make your calculations (see Appendix I, p. 172).

6. How would the apparent isotopic age of an igneous rock be affected if some of the daughter material had escaped from the rock? (This can happen if a rock is reheated during metamorphism). Explain and/or provide a simple mathematical illustration to justify your answer.

7. Speculate about why it may be difficult to obtain a meaningful isotopic age from a sedimentary rock.

IV. The geologic time scale

Gradually, using isotopic age determinations together with correlation and inferences about relative ages, absolute ages have been assigned to the fossil-based geologic time scale:

TABLE 2-4.
THE MODERN GEOLOGIC TIME SCALE

Melt glass from → *the Chicxulub crater, site of the meteorite impact that may have vanquished the dinosaurs, has been dated isotopically to almost exactly 65 million years -- the accepted age of the Cretaceous-Tertiary boundary.*

The geologic time scale is still being calibrated. As recently as 1993, the accepted age of the beginning of the Cambrian period was changed from 570 to 544 million years on the basis of new U-Pb dates from volcanic rocks in north-eastern Siberia . →

The new date decreases the length of the "Cambrian explosion" -- a time of rapid expansion in the diversity of life forms. (see Science, v. *261, p. 1293)*

EON	ERA	PERIOD	EPOCH	ABSOLUTE AGE
Phanerozoic	Cenozoic	Quaternary	Holocene	
				Began 10,000 yrs before present
			Pleistocene	
				1.64 million yrs $(1.64 \times 10^6$ yrs)
		Tertiary		
				65 million yrs $(6.50 \times 10^7$ yrs)
	Mesozoic	Cretaceous		
				145 million yrs
		Jurassic		
				208 million yrs
		Triassic		
				245 million yrs
	Paleozoic	Permian		
				290 million yrs
		Carboniferous		
		Pennsylvanian		(320 million yrs)
		Mississippian		362 million yrs
		Devonian		
				409 million yrs
		Silurian		
				439 million yrs
		Ordovician		
				510 million yrs
		Cambrian		
				544 million yrs
Precambrian	Proterozoic			
				2.50 billion yrs
	Archean			
Origin of Earth				4.55 billion yrs $(4.55 \times 10^9$ yrs)

FIGURE 2-4.
ABSOLUTE AGES
OF IGNEOUS
ROCKS EXPOSED
IN THE GRAND
CANYON
*The abbreviation
"my" stands for
millions of years*

A. Integrating relative and absolute ages
Isotopic ages for rocks of the Grand Canyon are shown in Figure 2-4. Use these to answer the following questions about the geologic history of the region.

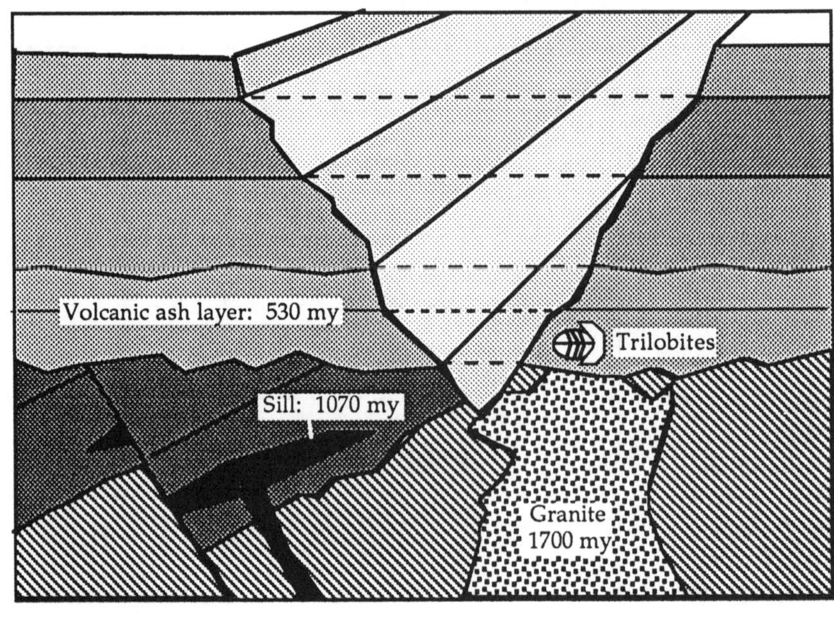

1. Were your inferences about the relative ages of the rocks in exercise I.B.2 consistent with the isotopic ages ? If not, determine why and correct your answers.

2. The chronology of two major igneous events could not be resolved using principles of relative dating. Which events are these and how do the isotopic ages resolve the ambiguity?

3. What is the range of possible absolute ages of the layer containing trilobites? How did you infer this? During which geologic period was layer probably deposited? (See Table 2-4 and Figure 2-3).

crust:
outermost layer of
Earth, 5-90 km thick

gneiss:
intensely meta-
morphosed rock with
prominent banding

*The oldest Moon rocks
yield isotopic ages
of 4.47 x 10^9 yrs.*

B. Age of the Earth
From the beginning, Earth's surface has been dynamic and changing. The face of the early Earth was so mobile, in fact, that no rocks of the original <u>crust</u> survive. The oldest rocks discovered on Earth so far are <u>gneisses</u> from northwestern Canada, dated isotopically at 3.964×10^9 (3.964 billion) yrs. The age of the Earth, however, is generally cited as 4.55×10^9 yrs (Table 2-4).

If there are no Earth rocks this old, what could have been dated to obtain this estimate? Hint: What are the only accessible, unaltered remnants from the earliest days of the Solar System? Review Section II.B., Laboratory 1.

V. Synthesis

A. Preview review
Return to the Lab Preview and complete any questions you were unable to answer before.

B. Detective work
Find an example of a geological, archeological, historical or legal investigation whose main purpose was to reconstruct events in the past. Describe the techniques used to determine the relative and absolute timing of the events. Cite at least 1 outside reference.

C. Geologic time analogy*
Isotopic dating has shown that the best understood part of the geologic record -- the fossil-rich Phanerozoic (Table 2-4) -- is only about 1/8 of the age of the Earth. Relative to human life spans, the vastness of geologic time is almost incomprehensible. It is helpful to think of analogies that express geologic time divisions as proportions of familiar quantities. You may have seen geologic time represented as a calendar year or a football field; develop your own proportional analogy illustrating:

> The age of the oldest surviving rocks on Earth
> The length of the Precambrian Eon
> The age of the bedrock in your area
> The extinction of the dinosaurs
> Your appearance on Earth

Hint: Set up equations of proportionality as you did in drawing the relative sizes of Earth and the Sun (Lab 1, Exercise **I.C.1**). Show all of your calculations.

---

* Based on an exercise described by Ritger, S. and Cummins, C., 1991, Journal of Geol. Education, v. 39

SUGGESTED REFERENCES AND RESOURCES

Time/ General:

 Gould, S., 1987, *Time's Arrow, Time's Cycle: Myth and Metaphor in the Discovery of Geological Time.* Cambridge, MA: Harvard University Press. 222 pp.

A scholarly account of the history of thinking about "Deep Time", centered on the theme of time as both cyclical and linear. The works of James Hutton are analyzed in a refreshingly irreverent and interesting way.

 Harland, W.B., Armstrong, R., Cox, A., Craig, L., Smith, A., & Smith, D., 1990. *A Geologic Time Scale 1989.* Cambridge: Cambridge University Press. 263 pp.

Unquestionably the most comprehensive source of information on the calibration of the geologic time scale.

 Thackeray, J. 1980. *The Age of the Earth.* London: Her Majesty's Stationery Office. 36 pp.

A brief but thorough and engagingly illustrated summary of the evidence for an Earth of great antiquity. Distributed by Cambridge University Press and available in the US through Pendragon House, Inc. 2595 East Bayshore Road. Palo Alto, CA 94303.

 Haq, B. and F. van Eysinga, 1987. Geologic Time Table (4th ed.) Amsterdam: Elsevier. Poster, 1 sheet.

Colorful poster with a wealth of time-related tables and figures: not only the standard geologic time scale but also significant paleontological divisions; names of correlative regional stratigraphic series; lunar stratigraphy; glacial and archeological time scales; as well as paleo-geographic and -oceanographic maps since the Jurassic. An invaluable laboratory resource.

The Grand Canyon:

 Elston, D., Billingsley, G. and Young, R., 1989. Geology of the Grand Canyon, Northern Arizona. Washington, DC: American Geophysical Union, 239 pp.

A detailed compendium of geologic and historical facts about the Canyon, published as a field guide for a down-river trip during the 28th International Geological Congress.

Tree-ring analysis:

 Butler, D. R., 1987. Teaching principles and applications of dendrogeomorphology. *Journal of Geological Education*, v. 35, p. 64-70.

A clear presentation of the logic and techniques of tree ring analysis, with emphasis on the use of tree rings to reconstruct the past frequency of avalanches and fires.

 Atwater, B. and Yamaguchi, D., 1991. Sudden, probably coseismic submergence of Holocene trees in coastal Washington State. *Geology*, v. 19, p. 706-709.

An interesting example of how tree-ring correlation has been used to date events in the recent past -- in this case, a major earthquake in the Pacific Northwest.

THE EARTH BENEATH OUR FEET

What controls the occurrence of volcanoes, earthquakes, and mountain ranges?

From a human perspective, great earthquakes and volcanic eruptions seem freakish and singular events. In a few seconds of ground shaking, apparently solid buildings are reduced to heaps of tangled debris. In a matter of weeks, a once placid mountain becomes a fuming menace. Then the shaking stops, the mountain becomes quiet again. People begin to rebuild, confident that such catastrophes will never be repeated. But if human lifetimes were centuries rather than decades long, we would recognize that these apparently extraordinary events are ordinary occurrences in the life of the Earth.

Although effective prediction of volcanic eruptions and particularly earthquakes is still more an art than a science, unmistakable patterns in the global distribution of these phenomena make it clear that they are not random, capricious occurrences. They reflect a planetary interior that is as mobile as the surface -- an interior intimately linked with the surface through a continuous process of crustal recycling. This process sets the tempo for an unending dance of the continents and sets the Earth apart from its planetary neighbors.

OBJECTIVES

- To discover what Earth's surface features reflect about its interior
- To explore patterns in the occurrence of earthquakes and volcanoes
- To consider why Earth's tectonic system appears to be unique
  among the other rocky planets and moons

OUTLINE

I. Continents and ocean basins
   Heights and depths
   Exceptions to the rules
   Volcanoes and earthquakes

II. Plate tectonics
   Plates and plate boundaries
   Deformation at plate boundaries
   Metamorphism at plate boundaries

III. Plate motions
   Insights from the ocean floor
   Rates of seafloor spreading
   The driving mechanism
   Hot spots

IV. Geologic hazards at plate boundaries
   Volcanic eruptions
   Volcanic rocks
   Earthquakes

LAB PREVIEW   Before coming to lab,
- Answer as many of the following questions as you can
- Read Part 2: "The Earth Beneath our Feet" (Chaps. 3-7) in *The Blue Planet*, or other readings assigned by your instructor

How much do you know already about the Earth beneath your feet?

Why does Earth have distinct continents and ocean basins?

What is the highest point on Earth's surface? What is the lowest point? What are the explanations for the extreme elevations of these places?

Why are volcanic eruptions in Hawaii generally less violent than those around the edges of the Pacific Ocean?

What causes an earthquake?

Why are earthquakes rare in the middle of the continental US?

How are the magnitude and epicenter of an earthquake determined?

EXERCISES

mode (statistical): the most common value of or range of values on a histogram

intrusive rock: igneous rock emplaced into the subsurface

extrusive rock: volcanic igneous rock, erupted onto Earth's surface

I. Continents and ocean basins

In Laboratory 1 (The *Earth in Space*), you found that the distributions of surface elevations on Earth and Venus, the most Earth-like planet, are fundamentally different. A histogram of surface elevations for Earth shows two well-defined maxima, while a similar plot for Venus shows only one (Fig. 1-6). Discovering the origin of Earth's distinctive surface topography is the first step toward understanding processes in the planet's interior.

A. Heights and depths

1. Review the distribution of surface elevations shown for Earth in Figure 1-6. What major surface features do the two peaks on the plot represent?

What is the most common range (<u>mode</u>) of elevations on the continents (the elevations associated with the tallest bar in the first peak on the plot)?

What is the most common range of depths in the oceans (the elevations associated with the tallest bar in the second peak on the plot)?

So what is the typical difference in elevation between continents and oceans?

2. Average elevation is not the only distinction between continents and ocean basins. Continental and oceanic rocks also have strikingly different compositions:

TABLE 3-1.
CONTRASTS BETWEEN CONTINENTS AND OCEAN BASINS

Feldspar, pyroxene, olivine, and other minerals are really groups of minerals with common crystal structures but varying compositions.

	Continents	Ocean basins
Typical rock type	Granite *Though made up of many different rock types, continents have an average composition comparable to that of this <u>intrusive</u> igneous rock, which consists of more than 65% silica (SiO_2)*	Basalt *Rocks on the ocean floor are much more uniform, made largely of this <u>extrusive</u> (volcanic) igneous rock with only about 50% silica*
Bulk density	2.75 g/cm^3	3.0 g/cm^3

Main minerals (in order of abundance)	Potassium & sodium feldspar
	  orthoclase: $KAlSi_3O_8$
	  albite: $NaAlSi_3O_8$
	Quartz SiO_2
	Biotite
	  $K_2(Mg,Fe)_2(OH_2)(AlSi_3O_{10})$
		Calcium feldspar
		  anorthite: $CaAl_2Si_2O_8$
		Pyroxene
		  augite: $Ca(Mg,Fe)(SiO_3)_2$
		Olivine
		  $(Mg,Fe)_2SiO_4$

Examine the hand specimens of granite and basalt and of the minerals in them. See if you can identify individual minerals in the rocks. (Crystals in the basalt may be too small to see without a hand lens or microscope). See also Appendix IV, Rock and mineral identification charts.

If you've forgotten the definition of density, review pp. 6-8. Lab 1.

asthenosphere: zone of weak, and flowing (but solid) rocks in Earth's upper mantle

isostasy: the state of flotational balance of the crust and uppermost mantle on the asthenosphere

FIGURE 3-1:
PRINCIPLE OF ISOSTASY
ρ = density
Δh = difference in elevations of continents & oceans
T_c ,T_o= thicknesses of continental and oceanic crust
T_{mc} ,T_{mo}= thickness of mantle between the base of continental or oceanic crust and the top of the asthenosphere

3. Compare the bulk density of Earth (Table 1-2, Lab 1) with the densities of continental and oceanic rocks. What does this suggest about the density of Earth's deeper interior? (See also Exercise I.D.5, Lab 1).

4. The fact that Earth's low-density granitic continents stand above the denser basaltic ocean basins suggests that both are "floating " on a still denser layer. The continental and oceanic rocks constitute only the outermost skin of the solid Earth, the crust. Continental crust is typically 30-60 km thick, while oceanic crust is only 5-10 km thick. Beneath the crust lies the denser mantle, made up mainly of the mineral olivine. Though solid, the mantle is hot enough to flow over long periods of time, just as glacial ice flows as a solid.

Part of the uppermost mantle, the <u>asthenosphere,</u> is particularly weak and will flow when subjected to uneven crustal loads, much as the water in a waterbed readjusts when you lie on it. Over time, this kind of readjustment in the mantle creates a state of balance, called <u>isostasy,</u> in which the total mass of rocks above the asthenosphere is the same from place to place.

Imagine 2 columns of rock -- one on a continent the other in the ocean -- extending from the Earth's surface to the top of the asthenosphere. At that depth (about 100 km below the seafloor), the total masses of the columns must be equal:

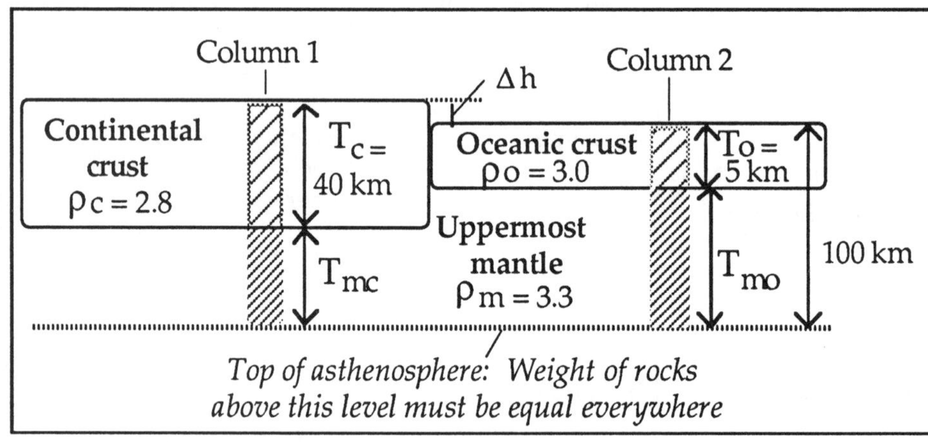

How much is Δh, the difference in elevation between continents and ocean basins that arises from their differences in density? To find out, we need mathematical expressions for the total mass of the rocks in the two columns. As a first step, write an equation for the mass of an object in terms of its density and volume:

$$\text{Mass} = \qquad\qquad\qquad (3\text{-}1)$$

The principle of isostasy is clearly illustrated by objects of different density floating in water. A piece of styrofoam, for example, floats high in the water, while a water-logged piece of wood just skims the surface.

Now write an expression for the total mass of Column 1. Let V_c = Volume of continental crust and V_{mc} = volume of mantle between the base of the continental crust and the asthenosphere. Then write a similar expression for Column 2.

Total mass of Column 1 =   +   (3-2)

Total mass of Column 2 =   +   (3-3)

If we make the columns square in map view, 1 km on a side, we can ignore their width and breadth and substitute their thicknesses (T_c, T_o, T_{mc}, T_{mo}) for their volumes. Rewrite your equations for the total masses making these substitutions.

Total mass of Column 1 =   +   (3-2a)

Total mass of Column 2 =   +   (3-3a)

The values for most quantities in these equations are shown in the figure above. Circle any quantities whose values are not known.

To constrain the values of these unknown quantities we need additional equations that contain them. One such equation relates the thicknesses of oceanic crust and underlying mantle to the depth of the asthenosphere. Refer to Figure 3-1 and complete the equation below.

$$T_o + \underline{\quad\quad} = 100 \text{ km} \qquad \text{Or, rearranging terms} \quad \underline{\quad\quad} = 100 - T_o \qquad (3\text{-}4)$$

If the ocean crust is 5 km thick, what is the actual value for this quantity?
$$T_{mo} = \underline{\quad\quad} \text{ km}$$

You now have values for all quantities in equations *(3-2a)* and *(3-3a)* except T_{mc}, the thickness of mantle between the continental crust and the asthenosphere. But we can find the value by setting equations *(3-2a)* and *(3-3a)* equal to each other (remember, masses of the 2 columns must be the same) and solving for T_{mc}:

Use the space above to verify equation (3-5) for yourself.

$$T_{mc} = [T_o \rho_o + (T_{mo}) \rho_m - T_c \rho_c]/\rho_m \qquad (3\text{-}5)$$

Use *(3-5)* and information given in Figure 3-1 to find the value of T_{mc}. (You can ignore the units on the density terms; they cancel each other out).
$$T_{mc} = \underline{\quad\quad} \text{ km}$$

Finally, complete the following equation relating the thicknesses of continental crust and underlying mantle to the depth of the asthenosphere and the difference in elevation between continents and oceans (refer to Fig. 3-1).

$$T_c + \underline{\quad} = 100 \text{ km} + \Delta h \qquad \text{Or, rearranging,} \quad \Delta h = T_c + \underline{\quad} - 100 \text{ km} \qquad (3\text{-}6)$$

Use *(3-6)* to solve for the difference in elevations between continents and oceans:
$$\Delta h = \underline{\quad\quad}$$

lithosphere: strong outer layer of the solid Earth, including the crust and uppermost mantle. Typically about 100 km thick.

Is this comparable to the value you determined previously (Exercise II.A.1)? You have shown that the different elevations of continents and oceans arise from their contrasting densities and the existence of a weak layer in Earth's mantle. The relatively rigid crust and mantle above this layer are together called the lithosphere. Later we will find that the contrasting strengths of the lithosphere and asthenosphere are of fundamental importance in Earth's tectonic system.

B. Exceptions to the rules

1. Some continental crust (crust that is granitic in composition) is actually below sea level. Using the surface elevation data for Earth in Figure 1-6, estimate what percent of Earth's surface area lies between 0 and 1000 m below sea level:

2. Study the ocean floor map in the lab (or Figure 6.13 in *The Blue Planet*), and describe where this shallowly submerged continental crust occurs.

continental shelves:
areas of shallowly submerged continental crust bordering the continents.

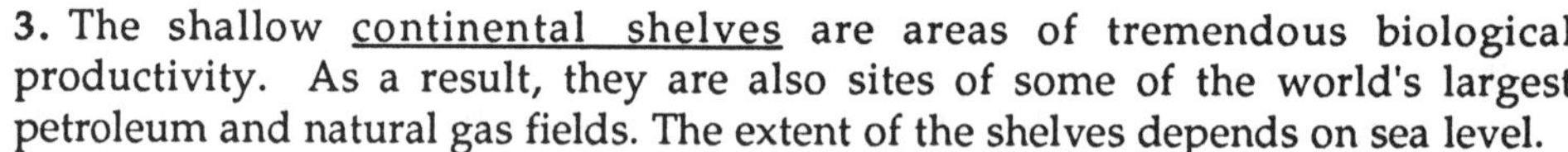

3. The shallow <u>continental shelves</u> are areas of tremendous biological productivity. As a result, they are also sites of some of the world's largest petroleum and natural gas fields. The extent of the shelves depends on sea level.

How would sea level be affected if the average density of oceanic crust were slightly lower or higher than the value in Figure 3-1? How would this affect the extent of the continental shelves? Explain your reasoning.

4. Some oceanic crust stands above the mean depth of the ocean floor. Using the surface elevation data for Earth in Figure 1-6, estimate the percent of Earth's surface area that lies between 1000 and 3000 m below sea level:

Look again at the ocean floor map and then describe where this relatively high-standing oceanic crust occurs.

What does the height of these parts of the ocean floor suggest about the density of the oceanic crust there? Explain.

5. A small percentage of the Earth's surface area lies well outside the averages. Using the surface elevation plot (Fig. 1-6) and the maps or other references provided, identify the highest continental area and deepest oceanic area.

Highest area on Earth's surface: Elevation:
Lowest area on Earth's surface: Depth:

earthquake **focus** (plural **foci**): the origin of an earthquake; point in the subsurface where fault slip begins

C. Volcanoes and earthquakes

1. Examine Figure 3-2, which shows the global distribution of earthquakes and active volcanoes (see also Figs. 5.19 and 6.9 in *The Blue Planet*). Describe their patterns of occurrence. (e.g., Do earthquakes and volcanoes tend to occur together? Are they clustered or randomly distributed? Where do the most explosive volcanoes occur? Where do the earthquakes with the deepest <u>foci</u> occur?)

2. Locate on Figure 3-2 the places you identified as Earth's highest and lowest points. Are these places characterized by seismic and volcanic activity? If so, describe the patterns of <u>seismicity</u> and/or volcanism in these areas. (Do they have clustered or scattered volcanoes and earthquakes? Deep or shallow earthquake foci?)

seismicity: earthquake activity

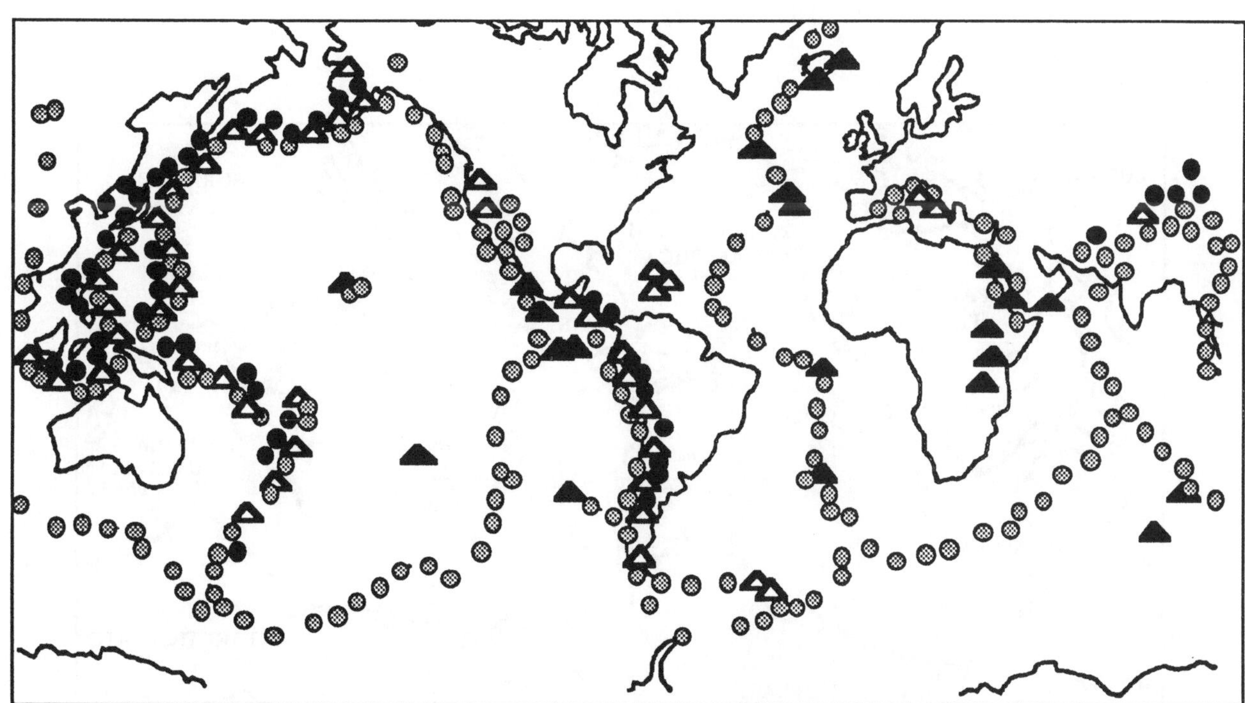

FIGURE 3-2. GLOBAL PATTERNS OF SEISMICITY AND VOLCANISM

◉ = Shallow-focus earthquakes (< 50 km) ● = Deep-focus earthquakes (50- 650 km)
△ = Explosive volcanoes ▲ = Non- explosive volcanoes

plate tectonics:
theory that Earth's lithosphere is a mosaic of moving pieces or plates

divergent boundary:
boundary at which plates move apart. In an ocean basin, a *mid-ocean ridge*

convergent boundary:
boundary at which plate move together via *subduction*

subduction:
return of dense ocean crust to the mantle

trench:
deep ocean trough at site of subduction

transform boundary:
boundary at which plates slide past each other

II. Plate tectonics

Recognition of such patterns of global volcanism and seismicity was an important step toward the development of a unifying theory of how the solid Earth works. The theory of <u>plate tectonics</u> is very young as scientific ideas go, having emerged only in the mid-1960's. The theory provides elegantly simple explanations for a wide range of phenomena including earthquakes, volcanoes and mountain ranges.

According to plate tectonics, the relatively narrow zones where most earthquakes occur mark boundaries between major pieces of the strong outer layer of the solid Earth, the lithosphere. There are a dozen major lithospheric pieces or plates (Fig. 3-3), each about 100 km thick. These plates move horizontally at speeds of 2 to 10 cm/yr -- about as fast as your fingernails grow. The plates are generally rigid but deform (change their shape) near their edges, where they interact with adjacent plates. Earthquakes are one consequence of this deformation.

There are three types of boundaries between tectonic plates. At <u>divergent</u> boundaries, adjacent plates move apart, and new lithosphere forms as magma wells up from Earth's interior. The Mid-Atlantic ridge is an example of a divergent boundary. At <u>convergent</u> boundaries, adjacent plates move toward each another and one (always oceanic) plunges beneath the other in a process called <u>subduction</u>. A deep oceanic <u>trench</u>, like the Marianas trench in the western Pacific, occurs where the two plates meet. Volcanism is also associated with these convergent boundaries because the downgoing plate and the mantle rocks above it undergo partial melting, generating magma that rises to the surface. The Cascade Mountains of the northwestern US, including Mt. St. Helens, are volcanoes associated with subduction. Finally, at <u>transform</u> boundaries, two adjacent plates simply slide past each other. California's San Andreas fault is a transform plate boundary.

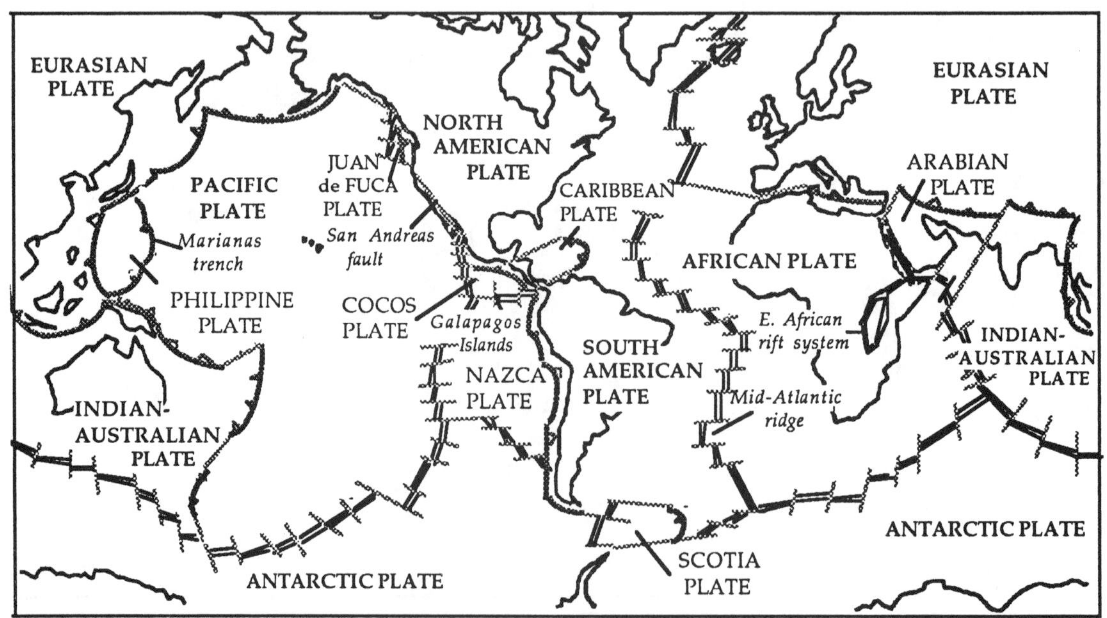

FIGURE 3-3. EARTH'S TECTONIC PLATES

= Divergent boundary   = Convergent boundary (teeth on upper plate)   = Transform boundary

A. Plate boundaries
Study Figure 3-3 (or Fig. 6.9 in *The Blue Planet*).

1. What type of plate boundary is nearest to where you live? Which plates are involved?

2. Do plate boundaries always coincide with coastlines?

Name 2 plates that include both continental and oceanic crust.

Are any plates entirely continental or entirely oceanic? If yes, list them.

3. Where do most divergent plate boundaries occur -- in oceanic or continental crust? Identify a notable exception to this generalization.

4. What type of boundary tends to be characterized by explosive volcanism? (Compare Figs. 3-2 and 3-3). Give a specific example.

What type of boundary tends to be characterized by non-explosive volcanism? Give a specific example.

Are there any areas of active volcanism that do not lie on or near plate boundaries? If so, give an example.

5. Do the places you identified as Earth's highest and lowest points coincide with plate boundaries? If so, what types of boundaries?

6. (Optional) Telephone the taped "Earthquake hotline" at the National Earthquake data center in Golden Colorado: (303) 273-8516. The recording lists all earthquakes of magnitude 5 and greater that have occurred around the world in the past 24 hours. Write down the locations of these earthquakes and determine whether or not they occurred on a plate boundary. If so, identify the type of plate boundary. (Note: This is not a toll-free telephone call).

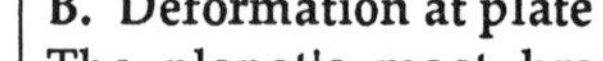

B. Deformation at plate boundaries
The planet's most breathtaking mountainscapes and deadliest earthquakes reflect rock deformation at plate boundaries --particularly convergent boundaries.

1. Look again at Figure 3-2 and compare the widths of the seismically active zones at the following convergent boundaries:
  • The Nazca-South American plate boundary on the W coast of South America
  • The boundary between the Indian-Australian and Eurasian plates NE of India
How might you explain the difference in the extent of rock deformation at these two boundaries?

stress:
unequal forces
applied in
different
directions

**brittle
deformation:**
change in shape by
fracturing and
frictional sliding

**ductile
deformation:**
change in shape by
flow rather than
fracturing

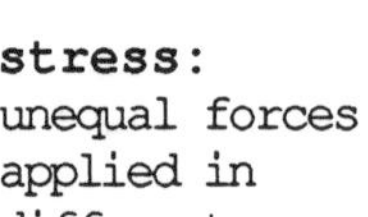

The widely distributed seismicity in the Himalayan region is typical of areas where two continental plates collide. The buoyant crust resists being subducted, and convergence between the plates must instead be accommodated by horizontal shortening and vertical thickening of the crust. The tremendous height of the Himalaya reflects such deformation of rock at the boundary between the Eurasian and Indian-Australian plates. Other great mountain belts, including the much older Appalachian chain, also developed by continent-continent collision.

Like many other substances, the way that rocks deform, or change their shape, under _stress_ depends largely on temperature.   At low temperatures (<300° C) typical of shallow depths (<15 km), rocks behave _brittlely_ -- that is, deformation occurs by formation of fractures (faults) and by frictional sliding on these fractures. Earthquakes occur when rocks deform brittlely. In contrast, at the higher temperatures typical of greater depths, rocks deform _ductilely_ -- though solid, they "flow". Deformation is distributed throughout the rock rather than concentrated along distinct fractures.

2. Explore the effect of temperature on material behavior by bending a piece of cold wax, then warming it for a minute in your hands and bending it again.

Play with the wax for a while. How does the rate at which you bend the wax affect the way it deforms?

Describe at least one other common material whose mechanical behavior is sensitive to temperature and/or rate of shape change.

folds:
wrinkles or buckles
in layered rock,
usually formed in
response to layer-
parallel shortening

joints:
fractures in rock
along which no slip
has occurred

slickensides:
lines or striae on
a fault surface,
recording the
direction of slip

veins:
mineral-filled
fractures. In some
veins, minerals
formed from magma;
in others they were
precipitated from
groundwater

metamorphism:
growth of new
minerals in rocks
in response to high
pressure and/or
temperature

porphyroblast:
crystal formed
during metamorphism

foliation:
planar surface in a
rock, formed by
alignment of platy
minerals like micas

shield:
core of a
continent, where
the oldest rocks
are found

3. Examine and sketch the specimens of deformed rock from plate boundary settings. Most of these structures can form at almost any scale -- so in a sense each represents a miniature mountain belt.

Which specimens experienced brittle deformation?   Ductile deformation? Explain.

C. Metamorphism at plate boundaries

In addition to being deformed, rocks commonly undergo mineralogical changes in response to the elevated temperatures and pressures they experience at plate boundaries. Such changes are called <u>metamorphism</u>. If metamorphism is not extreme, original sedimentary or igneous features may still be discernible. At higher degrees (grades) of metamorphism, all original features may be erased.

During metamorphism, large new crystals called <u>porphyroblasts</u> may grow in the rocks. The compositions of these crystals can be used to infer the temperatures and pressures at which metamorphism occurred. Garnet, used as a gemstone and in abrasives, is an example of a mineral that typically occurs as metamorphic porphyroblasts. If rocks undergo deformation at the same time as metamorphism, platy minerals like micas may grow in a preferred orientations, creating planar parting surfaces or <u>foliations</u> in the rock.

1. Earth's oldest rocks -- those exposed in the ancient <u>shield</u> area of continents -- have invariably experienced at least one and sometimes several episodes of metamorphism. Why is it not surprising that all of these ancient rocks have been altered by metamorphism? Think tectonically.

In Laboratory 2 (Earth in Time), you examined specimens of metamorphic rock types exposed in the deepest parts of the Grand Canyon.

2. Examine the specimens of metamorphic rocks and the 'parent' rocks, or protoliths, from which they formed (Table 3-2). Use a hand lens or binocular microscope for a closer look.

Can you recognize original sedimentary or igneous features in any of the specimens? If so, describe these features. Which specimens contain porphyroblasts? Which are foliated?

TABLE 3-2. METAMORPHIC ROCKS AND THEIR PROTOLITHS

Metamorphic rocks	Parent rock or protolith
Marble	Limestone
Quartzite	Sandstone
Slate	Shale (mudrock)
Phyllite	Shale
Schist	Shale
Gneiss	Igneous or sedimentary rocks
Amphibolite	Basalt
High-grade coal (anthracite)	Low-grade coal (lignite)

For centuries, sculptors including Michelangelo have preferred pure white "Carrera" marble, from a quarry in Tuscany by that name.

3. Because of their beauty, durability and capacity to split easily on foliation planes, metamorphic rocks are commonly used as building materials. Working in groups of 2 or 3, identify examples of metamorphic rocks in and on buildings on your campus and take careful note of their locations, then compile your "field" observations on a map of the campus.

To help you get started here are some typical uses of metamorphic rocks:
    Marble: Interior walls
    Slate: Roofs, flooring
    Gneiss: Exterior building stone
    Quartzite: Decorative stones around shrubs

III. Plate motions

We have now explored some of the evidence that Earth's lithosphere is broken into plates that are in constant relative motion. But what causes the plates to move? How fast do they move, and where have they been in the past?

A. Clues from the seafloor

Much of the evidence leading to plate tectonic theory came from defense-related mapping of the seafloor in the 1950's and 60's. Before then, almost nothing was known about the rocks beneath the deep ocean, and there was little reason to suspect that the geology of the ocean floor was much different from that of the continents. But the ocean floor turned out to be a wholly unfamiliar realm. The discovery of the global mid-ocean ridge system was the among the first surprising results of the seafloor mapping projects. Another surprise was the spatial pattern in the ages of ocean floor rocks.

1. Examine Figure 3-4, a map showing the age of oceanic crust around the world. Where does the oldest oceanic crust occur and what is its age?

How does this compare with the age of the oldest rocks found on the continents? (Review last page of exercises in Laboratory 2).

Clearly, something prevents ocean crust from reaching old age. What is it?

2. Study the ages of ocean floor rocks in the Atlantic. Where does the youngest oceanic crust occur? Where does the oldest occur?

Which part of the Atlantic opened first? Explain.

FIGURE 3-4:
AGE OF EARTH'S OCEANIC CRUST
Based on marine magnetic record. Lines bounding ocean crust of different ages represent magnetic <u>isochrons</u> (see next page).

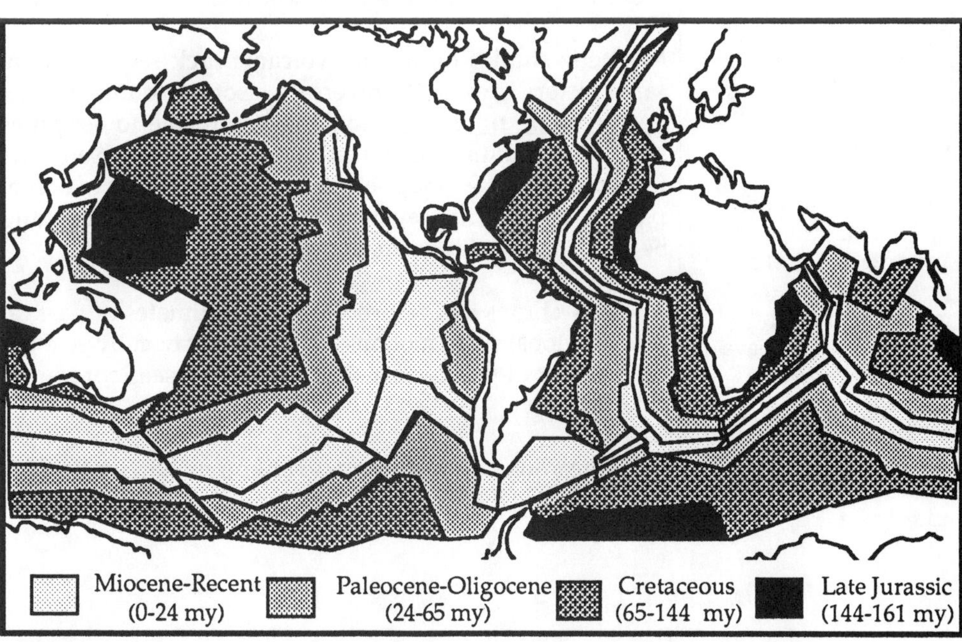

seafloor spreading:
process in which new oceanic lithosphere is formed by volcanic activity at mid-ocean ridges and then displaced progressively outward

Communities of strange creatures including giant tubes worms and albino crabs colonize the active parts of the mid-ocean ridge system. In frigid water reached by no sunlight, these organisms subsist on the heat and chemical nutrients vented from the submarine volcanoes.

polarity:
the sense of magnetization of a body -- i.e. the relative positions of its north and south poles

geomagnetic time scale:
time scale based on the history of reversals in the polarity of Earth's magnetic field. Absolute ages have been assigned to the reversals through isotopic dating of rocks formed close to the time of a reversal.

3. The increasing age of oceanic crust with distance from the mid-ocean ridges records the continuous creation of new lithosphere by basaltic volcanism at the ridges. As magma wells up and is extruded at a ridge, existing volcanic rock is displaced progressively outward. By this process of <u>seafloor spreading</u>, oceans grow and continents move apart.

Whether near a mid-ocean ridge or far away, all oceanic lithosphere was created by the process of seafloor spreading and is uniformly basaltic in composition. Yet mid-ocean ridges stand higher than the surrounding seafloor.

How would you expect newly-created oceanic lithosphere to differ from older lithosphere and how does this explain the ridges' height? Think isostatically.

B. Rates of seafloor spreading

Seafloor spreading was first proposed in the mid-1960's to explain puzzling patterns on new magnetic maps of the seafloor. The maps showed prominent "stripes" that are parallel to and symmetrical about the mid-ocean ridges. The stripes represent alternating bands of rock whose iron-bearing minerals have either the same <u>polarity</u> as Earth's present magnetic field or the opposite polarity. The existence of oppositely magnetized rocks indicates that Earth's magnetic field has reversed its polarity in the past. The symmetry of the magnetic stripes on either side of the mid-ocean ridges suggests that rocks form at the ridges, "freeze in" the existing magnetic field, and then are split apart and moved away from the ridges as new rocks are created by continued volcanism.

Magnetic studies of layered volcanic rock sequences on land soon revealed the same history of polarity reversals. Isotopic dating of these sequences created the <u>geomagnetic time scale</u>, making it possible to assign ages to ocean floor rocks simply by counting how many normal and reverse polarity stripes separate them from the ridge. On maps of the ocean floor, areas of ocean crust corresponding to a particular polarity reversal are marked by lines called <u>magnetic isochrons</u>. Several such isochrons are shown in Figure 3-4.

These isochrons can be used to make estimates of past rates of seafloor spreading and of global plate motions, which are driven by seafloor spreading. If Δx is the map width of oceanic crust produced by seafloor spreading in some interval of time Δt, then:

$$\text{Spreading rate (e.g. in cm/yr or km/million yrs)} = \Delta x\,/\,\Delta t \qquad (3\text{-}7)$$

**magnetic
isochron:**
line on an ocean
floor map joining
rocks formed during
a particular
magnetic reversal.
*[Greek: iso = same
+ kronos = time]*

1. Use Figure 3-4 to determine the rate of seafloor spreading on the Mid-Atlantic Ridge at the latitude of southern Florida, for the period 0 to 65 my ago.

To do this, measure the map width of ocean crust between the 65 million year isochrons on either side of the spreading ridge. Be sure to make your measurements perpendicular to the ridge. The scale of the map at low latitudes is 1: 320,000,000 (1 cm = 3200 km). Use equation *3-7a* to determine your answer in cm/yr. Show all your work.

$$\frac{\text{Map width of ocean crust (cm)}}{\text{Time period in years}} \times \text{Map scaling factor} = \text{Spreading rate in cm/yr} \qquad (3\text{-}7a)$$

2. Now determine the rate of spreading for the same time period for the ridge that marks the western boundary of the Cocos plate in the Pacific basin (Fig. 3-3). In this case, because the early Cenozoic ocean crust east of the ridge has already been subducted, measure only the map width of ocean crust from the active ridge westward to the 65 million year isochron, then double that value to get the full spreading rate. Show all you work.

3. What might account for the differences in spreading rates in the Atlantic and Pacific Oceans? Hint: How are the margins of these two ocean basins different?

**paleogeographic
map:**
map showing
configurations of
oceans and land
masses at times in
the past

4. Time travel
Magnetic isochrons also allow detailed reconstructions of past plate positions. Place a transparent overlay on Figure 3-4, and trace a) the coastal outline of South America; and b) the 24 my isochron *west* of the mid-Atlantic ridge. Then move the overlay so that your tracing of the isochron coincides as closely as possible with the same isochron east of the ridge (remember, 24 million years ago, both isochrons lay at the spreading ridge). Now trace the coastal outline of Africa. You will have created a map showing the size of the Atlantic Ocean and the relative positions of South America and Africa 24 million years ago.

Using the same procedure, make a paleogeographic map showing the relative positions of Africa and South America 65 million years ago.

convection: gravitationally driven overturn of a heated material

When parent isotopes like uranium, thorium, rubidium and potassium decay to daughter products, they release heat energy. This is the energy source for plate tectonics -- and one reason nuclear waste is so difficult to store safely.

FIGURE 3-5. EARTH'S INTERIOR AND THE PLATE TECTONIC SYSTEM Bold arrows illustrate whole-mantle convection. Heights of volcanoes are greatly exaggerated.

C. The driving mechanism

What causes the plates to diverge, converge and collide? Earth's tectonic system is an expression of the planet's thermal budget, a delicate balance between the rate of heat generation by radioactive decay in Earth's interior and the rate of heat loss at Earth's surface.

The driving force for plate movement is thermal <u>convection,</u> or overturn, of the mantle. Convection occurs when a material is heated from below or within, then expands and rises due its increased buoyancy, and finally cools again and sinks. Convection occurs not only in the deep interior of the earth but also in the atmosphere, in the oceans, and in boiling pots of soup. Bear in mind, though, that the mantle is not liquid but solid rock that 'flows' over very long periods time.

Spreading ridges (divergent margins) are presumed to coincide with rising or upwelling areas of the convecting mantle, and subduction zones (convergent margins) with downwelling areas (Fig. 3-5). It is not yet clear whether all of Earth's mantle, or only the upper part, is involved in the convective overturn that governs plate motion.

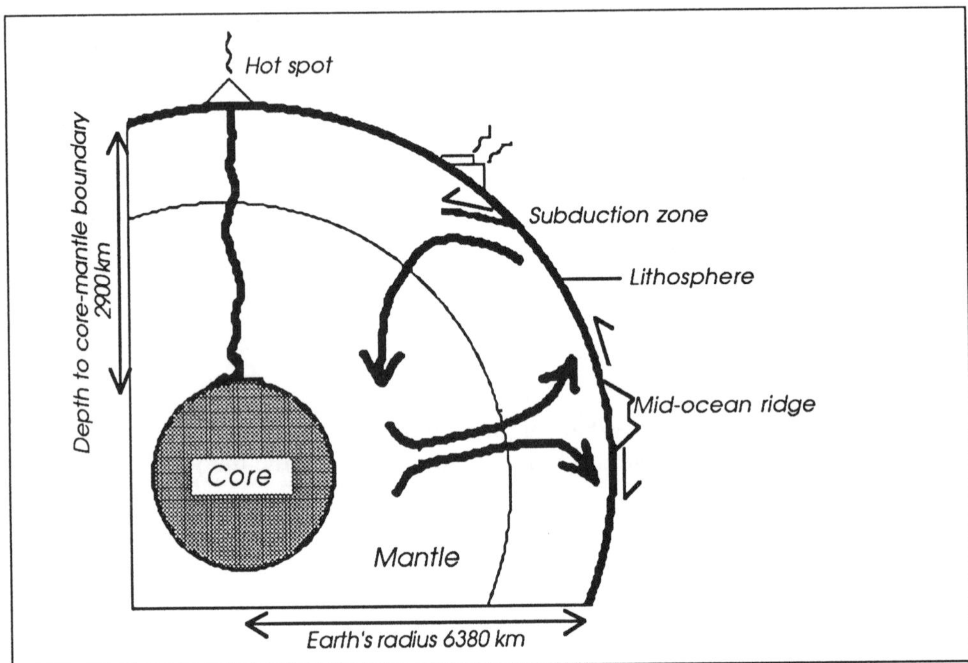

1. You can generate a miniature convective system by putting about 20 ml of dark corn syrup and 20 ml of light corn syrup in a small beaker and heating it in a water bath. Keep the water level no higher than the interface between the two types of syrup. One of the 2 types of syrup will initially be denser and sink to the bottom. As it is warmed, it will expand and become less dense than the cooler syrup above it.

Describe what you observe. Is there more than one area of upwelling and downwelling? Does the style of convection change as heating continues?

2. Mantle convection alone is not enough to establish a plate tectonic system, however. The existence of a thin, cold, rigid lithosphere from which to form plates is also a prerequisite. Subduction cannot occur unless parts of the lithosphere are cold and dense enough to sink into the mantle. It appears that Earth is the only body in the Solar System with a well-defined plate tectonic system, and Earth was probably too hot during its first half billion years for rigid lithospheric plates to exist.

The Moon has a 'one-plate' lithosphere estimated to be 700 km thick, about four-tenths of its 1700 km radius. The ratio of lithospheric thickness to planetary radius is thought to be comparable for Mars. What is the value of this ratio for Earth? Earth's lithosphere is about 100 km thick. For its radius, see Figure 3-5.

Because they have large surface areas relative to their volumes, small animals lose heat more quickly than large ones. So to maintain a constant body temperature, smaller animals must have higher metabolic rates.

3. Development of a thick lithosphere indicates that a planet or moon has lost much of its internal heat. Heat loss is proportional to surface area and small bodies like the Moon have relatively large surface areas compared to their volumes. This is because the surface area (A) of a sphere of radius r is $4 \pi r^2$ while the volume (V) is $4/3 \pi r^3$.

What is the general expression for the ratio of surface area to volume (A/V) for a sphere in terms of its radius, r?   **A/V**   =   __________   (3-8)

Determine the relative values of this quantity for Earth, Mercury, Venus, Mars and the Moon. For simplicity, first normalize the radii of these bodies relative to Earth's radius (i.e. divide their radii by Earth's radius) and enter these ratios (r') in Table 3-2. Then enter the relative area/volume values.

On the basis of the area/volume ratios, indicate the order in which you would expect these bodies to have cooled. Is this consistent with the relative ages of the planetary surfaces you determined in Laboratory 1 (Synthesis. p. 18)?

TABLE 3-3:
RELATIVE SURFACE AREA/ VOLUME RATIOS FOR THE ROCKY PLANETS

	Radius (km)	Radius relative to Earth's (r')	A/V=3/r'	Cooling order (1 = first)
Earth	6380	1.00	3	
Moon	1700			
Mercury	2440			
Venus	6052			
Mars	3398			

4. Which planet would you expect to lose heat at about the same rate as Earth?

This planet, however, seems to lack a rigid lithosphere like Earth's. Given what you know about conditions at the surface of this planet, how can you explain this?

residence time: average length of time a material remains in a particular setting

5. On Earth, plate tectonics involves the continuous creation and destruction of oceanic lithosphere. How many times has this cycle occurred? Assuming that Earth first achieved the thermal conditions for plate tectonics about 3900 million years ago and that the age of the oldest existing seafloor (Fig. 3-4) represents the replacement time ("residence time") for oceanic crust, estimate how many times in Earth's history the ocean floor has been replaced. Show your calculations.

hot spot:
point on Earth's surface above an isolated column of unusually hot, buoyantly rising mantle rock. Magma generated as this rock nears the surface causes volcanic activity like that in Hawaii.

seamount:
an isolated submarine volcanic mountain, typically formed above a mantle hot spot

D. Hot spots

Mid-ocean ridges, marking convective upwellings of the mantle, are the sites where Earth vents most of its internal heat. However, isolated columns ("plumes") of hot mantle rock, also transport heat from Earth's interior (Fig. 3-5). A point on Earth's surface above a mantle plume is called a <u>hot spot</u>. The high-temperature mantle at hot spots partly melts the base of the lithosphere, giving rise to volcanism. Hawaii and Yellowstone are both sites of hot spot volcanism.

Unlike the wandering plates, mantle plumes are thought to be stationary relative to Earth's deep interior (they apparently originate close to the core-mantle boundary) and so provide a fixed frame of reference by which to gauge the motions of the lithospheric plates. As a plate moves over the plume, a chain of volcanoes is produced. At a particular time, the only active volcano is the one immediately above the plume. Volcanic hot spot chains or "trails" are easiest to recognize in ocean basins since the ocean basins are geologically and topographically simpler than the continents.

1. You can simulate the development of a volcanic hot spot trail using a marking pen as a mantle plume and a piece of transparent film or Plexiglas as a lithospheric plate. Mark a 'north' arrow on the Plexiglas, then hold the marking pen vertical and stationary, with the tip pointing up. Move the Plexiglas a short distance in a 'northerly' direction over the pen so that it leaves a continuous line. Then change the direction of plate motion to northwesterly. Make a simple map of your "volcanic chain" here:

On your map, label the locations of: 1) the oldest volcano; 2) the youngest volcano and 3) the point recording the time when the direction of plate motion changed.

2. Many of the chains of islands and submarine <u>seamounts</u> in the Pacific ocean are hot spot trails. If one is available, study the map showing the topography of the Pacific ocean floor and locate the Emperor seamount chain on the map. What group of islands does the Emperor chain appear to be continuous with?

In light of your simple experiment above, describe how this seamount/island chain formed.

3. The distance from the northernmost Emperor seamount, formed 79 million yrs ago, to the bend in the Hawaii-Emperor chain is about 3800 km. It is another 3600 km from the bend to the island of Hawaii, where volcanism occurs today.

Using this information, determine how fast has the Pacific plate been moving since the Hawaii hotspot has been active. How does this compare with your earlier estimate of plate velocity in the Pacific based on rates of seafloor spreading (Exercise II.B.2)? Show your calculations.

IV. Geologic hazards at plate boundaries

The 1980s and 1990s have brought the dangers of life near plate boundaries into the public consciousness. Since 1980, volcanic eruptions at Mt. St. Helens (Washington State), El Chichon (Mexico), Nevada del Ruiz (Columbia), Mt. Unzen (Japan), and Mt. Pinatubo (Philippines) have killed thousands and left many more homeless. Devastating earthquakes in Mexico City, Armenia, Iran, Egypt and California have exacted an even larger toll. What can be done to reduce human suffering in tectonically active areas? The first step is to understand which settings pose the greatest dangers.

A. Volcanic eruptions

1. Refer to the global plate tectonic map (Fig. 3-3) and a world atlas if necessary. What tectonic setting is common to Mt. St. Helens, El Chichon, Nevada del Ruiz, Mt. Unzen and Mt. Pinatubo?

2. Hawaii and Iceland are areas of almost continuous volcanic activity, yet few volcano-related deaths or injuries occur in these places. What are the tectonic settings of these two groups of islands? (See Figs. 3-3 and 3-4).

pyroclastic material:
volcanic debris thrown into the air during an explosive eruption; ranges in size from fine dust (ash) to >10 m blocks (bombs). Also called *tephra*.

stratovolcano:
volcano built from alternating layers of lava and ash

viscosity:
degree to which a fluid resists flow

mafic lava:
low-viscosity, low silica lava

silicic lava:
high-viscosity, high silica lava

The potential hazards associated with volcanism are clearly related to tectonic setting. Volcanoes form in 3 principal tectonic environments: near convergent plate boundaries, at divergent plate boundaries, and above hotspots. In general, eruptions at divergent margins and hotspots are almost continuous, non-explosive, and rarely dangerous to anyone taking reasonable precautions. They produce lava flows and sometimes lava fountains 300-500 m high. The flows gradually build broad, gently sloping <u>shield volcanoes</u> like those in Hawaii (or on Mars --see Lab 1). In contrast, eruptions near convergent plate boundaries are intermittent and commonly explosive, producing dangerous airfalls and over-ground flows of <u>pyroclastic</u> material. Pyroclastic flows move over the ground surface at tremendous speeds as glowing hot avalanches, as mudflows, or as directed lateral blasts from a collapsing volcano. Successive eruptions at convergent boundaries build steep <u>stratovolcanoes</u>, made of alternating layers of lava and ash.

The explosivity of a volcano is related to the <u>viscosity</u> (stickiness) of the lava it produces. The <u>mafic</u> or basaltic (< 53% SiO_2) lavas associated with hot spots and divergent plate margins have low viscosities. Gases dissolved in mafic lavas can readily escape, and the lava flows quietly and non-explosively. The more <u>silicic</u> or rhyolitic (53-78% SiO_2) lavas characteristic of subduction-related volcanoes, in contrast, have high viscosities. Expanding gas bubbles become trapped, and pressures build until violent eruptions occur.

3. Refer again to the Figure 3-3. Based on their tectonic settings, what type of eruptions (explosive or not) would you expect in these volcanically active areas:

<u>Location</u>	<u>Tectonic setting</u>	<u>Expected eruption type</u>
Italy		
Galapagos Islands		
Indonesia		

4. Of all volcanic hazards at convergent margins, pyroclastic flows cause the greatest loss of life because they are hot (200 - 1100 °C) and very fast-moving

On the day of the main Mt. St. Helens eruption, seismometers 3.5 miles from the blast origin stopped transmitting 77 seconds after the eruption, when they were overridden by a dense pyroclastic blast cloud. Based on this information, how fast (in miles per hour) was the blast cloud moving?

5. Although the 1980 eruption of Mt. St. Helens eruption was impressive, the volume of material ejected from the volcano was actually small in comparison to other historic eruptions at subduction-related volcanoes. Table 3-3 lists the estimated volumes of ejecta thrown out during selected historic eruptions.

The 1815 eruption of *Tambora* in Indonesia was the largest in historical times. The eruption was heard at distances of 2600 km, and there was complete darkness for two days within a 20 km radius of the volcano. More than 90,000 people perished in the pyroclastic flows and <u>tsunami</u> (seismic sea wave) that followed the eruption. Gases and fine ash injected into the atmosphere reflected incoming sunlight, causing such dramatic global cooling that 1816 became known as "the year without a summer". Seventy years later, the eruption of another Indonesian volcano, *Krakatau*, similarly influenced global weather patterns.

The eruption of Italy's *Mt. Vesuvius* in 79 A.D. buried (and preserved) the city of Pompeii under a blanket of ash.

Mt. Mazama, like Mt. St. Helens, was one of the volcanoes in the Cascade chain. Its eruption in about 5000 B.C., which formed Crater Lake, may have been witnessed by early Native Americans.

tsunami:
gigantic wave caused by a submarine earthquake. (Seismic activity often accompanies volcanic eruptions) *[Japanese: "port wave"]*

Laura Ingalls Wilder's book The Long Winter *describes one of the harsh winters that followed the 1883 eruption of Krakatau.*

TABLE 3-4:
VOLUME OF EJECTA
FROM HISTORIC
ERUPTIONS

The thick ash deposits on the flanks of Mt. Pinatubo will pose mudslide hazards long after the volcano becomes quiet again.

Year	Eruption	Volume of ejecta (km^3)
5000 BC	Mt. Mazama, Oregon	41
79 AD	Vesuvius, Italy	2.5
1707	Mt. Fuji, Japan	2
1815	Tambora, Indonesia	30
1842-57	Mt. St. Helens	1
1883	Krakatau, Indonesia	18
1912	Mt. Katmai, Alaska	12
1956	Bezymianny, Kamchatka, USSR	1
1980	Mt. St. Helens	0.5
1991	Mt. Pinatubo, Philippines	5

Using the data in Table 3-4, make a bar graph showing the volume of material ejected (vertical axis) in the listed eruptions (horizontal axis).

Which appears to have the potential to be more destructive: A convergent-margin volcano like Mt. St. Helens that erupts about once every century or one that erupts at much longer intervals? Why do you think this is so?

tuff:
rock formed from
compacted volcanic
ash. If ash was
still hot when it
accumulated, tuff
may be fused or
'welded'.

*The proposed nuclear
waste repository at
Yucca Mountain, NV
(Ch. 8) is in a welded
tuff, one of the least
permeable rock types*

obsidian:
volcanic glass

pumice:
rock formed from
gas-rich, glassy
volcanic "foam".
Pumice may have so
much air space that
it will float!

pahoehoe:
lava with a "ropy"
surface texture.
Typically formed in
thin basalt flows.

aa:
lava with a blocky
texture. Formed
when the surface of
basaltic lava has
nearly solidified
but flow continues.

B. Volcanic rocks

Examine the hand specimens of volcanic rocks. Describe and sketch them, then discuss what their textures reflect about their mode of formation. Which consist of pyroclastic material from explosive eruptions? Which are flows from quieter eruptions? Which cooled most quickly? If information on the origins of the specimens is available, identify the tectonic setting from which they came.

Terms describing volcanic textures are listed at left; terms describing volcanic compositions are given below.

porphyry:
igneous rock with
two sizes of
crystals: Large
ones formed slowly
in a chamber under
a volcano and finer
ones formed by
rapid cooling of
lava erupted at the
surface.

basalt:
low-silica (mafic)
volcanic rock, consist-
ing mainly of Ca- and
Na-rich feldspar and
pyroxenes. Typical of
divergent plate boun-
daries and oceanic hot
spots (and the lunar
lowlands). Intrusive
equivalent: **gabbro.**

rhyolite:
high-silica (silicic)
volcanic rock consist-
ing mainly of K- and
Na-rich feldspar,
quartz, and mica.
Common at convergent
plate boundaries and
continental hotspots.
Intrusive equivalent:
granite.

andesite:
volcanic rock with
composition
intermediate between
that of basalt and
rhyolite. Most common
rock type at
convergent plate
boundaries. Intrusive
equivalent: **diorite.**

P-wave: fastest type of seismic wave and the first to be recorded on seismograms (**P = Primary**)

S-wave: second type of seismic wave to be recorded on seismograms (**S = Secondary**).

surface wave: seismic wave that travels only along surfaces between materials of contrasting density

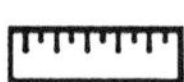 

epicenter: Point on the surface of the earth above the focus of an earthquake.

Epicentral locations are now determined by computers minutes after earthquakes occur, using P-S lag times from hundreds of seismic stations

C. Earthquakes

Earthquakes are the other major hazard associated with tectonic plate boundaries. Contrary to depictions in film and folklore, deaths and injuries caused by earthquakes are not related to the opening of great chasms in the earth. Instead, most injuries are related to the collapse of buildings due to ground shaking. This shaking is caused by the passage of vibrations generated at the point where fault slip begins (the focus). These vibrations travel as waves outward in all directions from the focus. Seismic waves generated by moderate to large earthquakes can be detected by seismometers around the world.

There are three main types of seismic wave energy generated by earthquakes, and each travels through the earth at a different speed. P-waves are the fastest waves, traveling at about 6.0 km/sec through the Earth's crust. S-waves are a bit slower, at 3.5 km/sec. Surface waves are still slower but are associated with the greatest ground shaking. Persons who have experienced major earthquakes sometimes describe having been able to detect the distinct arrivals of the three different wave types.

Because P-waves travel faster than S-waves, they arrive sooner at any given distance from the origin of the earthquake. The greater the distance from the origin, the greater the lag time between the arrival of the P- and S-waves. This fact is the key to locating the epicenter of an earthquake -- the location on Earth's surface directly above the focus.

1. Locating earthquake epicenters

Determine the epicenter of the earthquake assigned to your group, following the instructions below. Enter the quake number (1 or 2) in the caption of Table 3-5.

- From the seismic records of the quake (Fig. 3-6), determine the arrival times of P and S waves at the 3 stations, then calculate the lag time between P and S arrivals. Record this information in Table 3-5.

- Convert the lag times to distances from the epicenter using the lowermost curve in Figure 3-7.

- Using a compass --or simply a piece of string and a pen -- draw on the North American map (Fig. 3-8) three circles (or arcs) centered on the three stations, with radii equal to the distance from the epicenter.

Where is the epicenter? Is it near a plate boundary? If so, what type?

TABLE 3-5.
SEISMIC DATA FOR EARTHQUAKE #__

STATION	Time of P arrival (min)	Time of S arrival (min)	P-S lag time (min)	Distance from epicenter (km)
A				
B				
C				

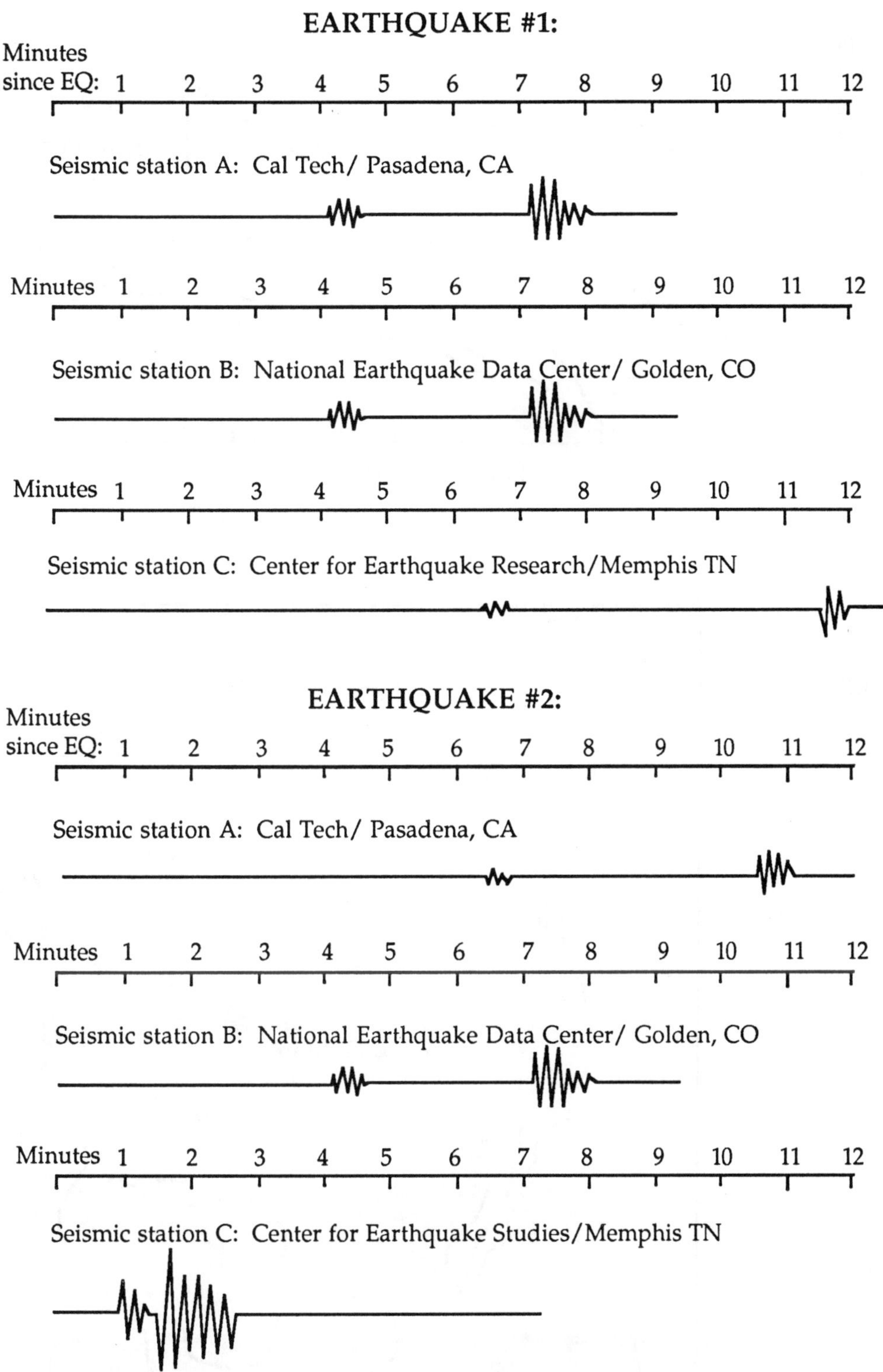

FIGURE 3-6. SEISMIC RECORDS OF TWO EARTHQUAKES AT THREE STATIONS
Surface wave arrivals not shown

FIGURE 3-7.
DIFFERENCE IN
ARRIVAL TIMES OF
P- AND S-WAVES
AS A FUNCTION OF
DISTANCE FROM
EARTHQUAKE
EPICENTER

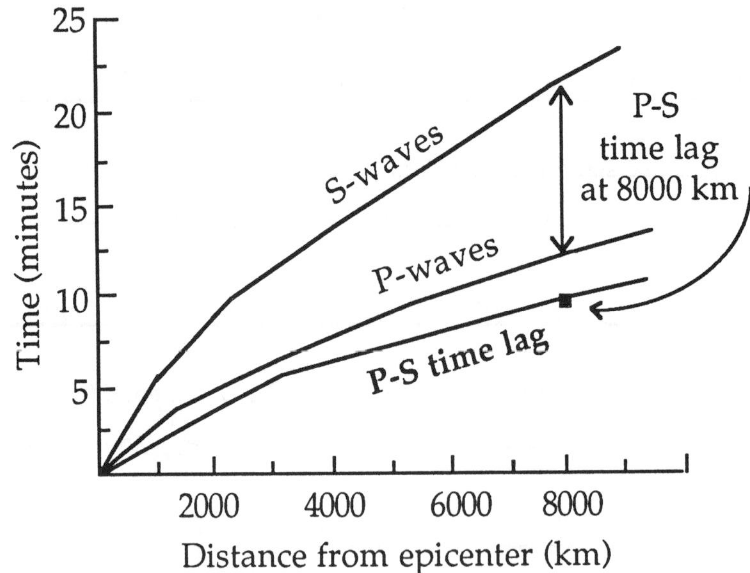

FIGURE 3-8.
MAP OF
NORTH AMERICA
(FOR USE IN
DETERMINATION
OF EARTHQUAKE
EPICENTERS)

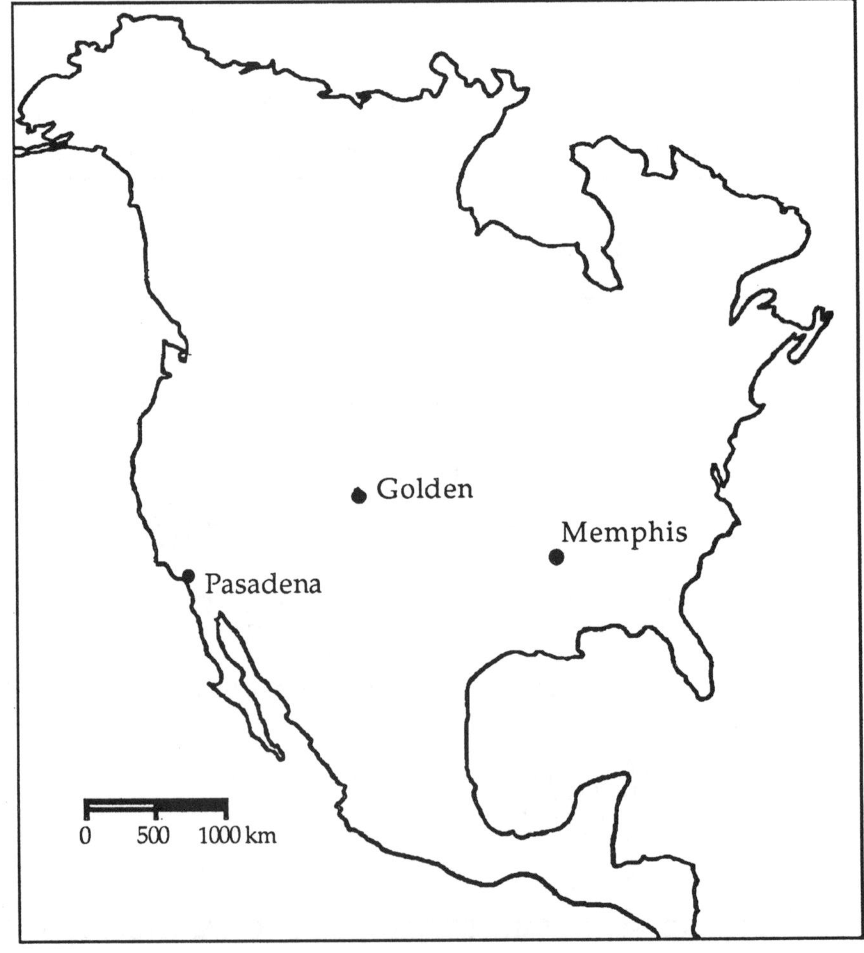

magnitude:
quantitative description of the size of an earthquake

Richter scale:
most commonly used earthquake magnitude scale, based on P-wave amplitudes

logarithmic scale:
measurement scale in which each unit represents an order-of-magnitude difference in value

 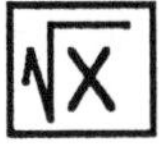

2. Assigning earthquake magnitudes

The <u>magnitude</u> of an earthquake is a description of the energy it releases. The most common magnitude scale, the <u>Richter scale,</u> is based on the height, or amplitude, of the largest arrival recorded on a seismogram, adjusted for distance from the epicenter. Wave amplitude is directly related to the total area of rupture (slip) on a fault, which is one factor governing the energy released in an earthquake.

Because earthquake magnitudes vary over such a wide range, the Richter scale is not linear but <u>logarithmic</u>. Each unit increase on the Richter scale corresponds to a ten-fold increase in P-wave amplitude and rupture area, and a *thirty*-fold increase in seismic energy. If a Richter magnitude 1 earthquake is assigned an energy value of 1, then the general relationship between earthquake magnitude (M) and energy (E) is:

$$E = 30^{M-1} \tag{3-9}$$

a) Using equation *3-9*, make a plot of earthquake energy release (*E*, vertical axis) vs. magnitude from 1 to 9 (*M*, horizontal axis).

b) Figure 3-9 shows simplified records for 5 seismic events, all recorded at a distance of 100 km from the epicenter. Determine the magnitudes of the earthquakes represented by comparing the amplitude of their first P-wave arrival with the P-wave amplitude on the record for the devastating 1988 magnitude 7 earthquake in Armenia. Remember that a 10-fold increase in P-wave amplitude (*A*) corresponds to a unit increase in magnitude, so if a magnitude 7 earthquake has registers a P-wave amplitude of 1 mm, then:

$$A\,10^{M-7} \qquad \text{(where A is in mm).} \tag{3-10}$$

Solving for M, this equation becomes

$$M = \log(A, mm) + 7 \tag{3-11}$$

Enter the P-wave amplitudes and magnitudes in the table below then add these earthquakes to your plot of magnitude vs. energy release.

TABLE 3-6.
MAGNITUDES OF SELECTED MAJOR EARTHQUAKES

EVENT	MAXIMUM AMPLITUDE	MAGNITUDE
1988: Armenia	1.00 mm	7.0
1989: Loma Prieta, CA ("World Series" earthquake)		
1976: Tangshan, China		
1964: Southern Alaska		

Because the seismic signature of explosions differs from that of earthquakes, underground nuclear tests can be detected by seismic monitoring.

The causes of the New Madrid and other intraplate earthquakes are poorly understood.

FIGURE 3-9.
SEISMOGRAMS FOR MAJOR HISTORIC EARTHQUAKES (FOR USE IN DETERMINATION OF EARTHQUAKE MAGNITUDES)

c) The seismic record for the explosion of a 1 megaton nuclear bomb is shown at the bottom of Figure 3-9. How does it differ from those of natural earthquakes? Determine its Richter magnitude and add it to your magnitude vs. energy plot.

d) Though uncommon, earthquakes sometimes do occur far from plate boundaries. In fact, some of the largest historical earthquakes in North America occurred in 1811-1812 in SE Missouri, near the small town of New Madrid. The magnitudes of these earthquakes are estimated to have been 8.0 to 8.3. Add the New Madrid earthquakes to your plot.

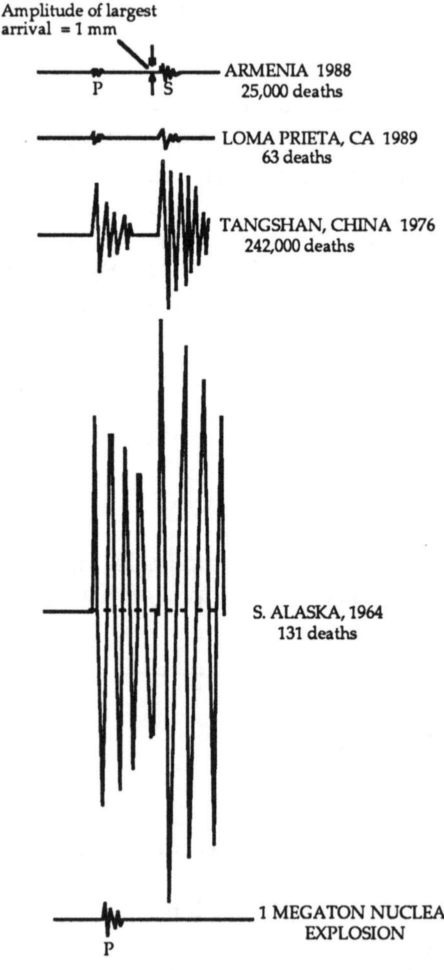

V. Synthesis

A. Preview review
Return to the Lab Preview and complete any questions you were unable to answer before.

B. Small worlds
The properties of the minerals that make up Earth's crust reflect the systematic arrangement of atoms within them. In micas like muscovite and biotite, for example, atoms are arranged in layers, causing these minerals to break or cleave into thin sheets. The atomic structures of minerals can be succinctly described by a few simple symmetry operations -- geometric rules for the repetition of patterns in space. These operations include translation, rotation, reflection, and glide (a combination of reflection and translation). (See your instructor and Ch. 4 in *The Blue Planet*). The same symmetry operations can be seen in spatially repeating patterns in the everyday world -- for example, in wallpaper or floor tiling. Describe and/or illustrate at least one such pattern and determine the symmetry operations that generate it. In other words, show the locations of all rotation axes, mirror planes, and glide planes and also the boundaries of the smallest unit (the 'unit cell') that can produce the pattern by translation.

C. Land and sea
Complete the following table summarizing the differences between continents and oceans.

Characteristic	Continents	Ocean basins
Composition:	Granitic on average, but varied	Basaltic, globally uniform
Density:		
Age of oldest crust:		
Type of plate boundary at which highest elevations occur:		
Other distinctions:		

D. Living with a living planet
Write a short essay on one of the following topics:

- If you have ever experienced an earthquake, write a short description it. Where were you? What did you feel? How did you and others behave?

- If you haven't experienced an earthquake, interview someone who has and ask them to respond to the same questions.

- Plate tectonic theory has made it clear that devastating earthquakes and volcanic eruptions are inevitable in some tectonically active areas. In light of the costs associated with disaster relief, do you think that governments should act to restrict habitation of areas where the potential for natural hazards is high? Write a short statement of your views on how humans can best cope with the planet's tectonic upheavals.

SUGGESTED REFERENCES AND RESOURCES

Tectonics/ General:

Earth magazine, Kalmbach Publishing Co., Waukesha, Wisconsin.

In additional to frequent feature articles on plate tectonic themes, each issue includes a summary of recent seismic and volcanic events around the world. Subscriptions are a very reasonable $20/year (6 issues). Kalmbach Publishing, 21027 Crossroads Circle, PO Box 1612, Waukesha, WI 53187. Telephone: 414/796-8776.

National Geographic maps of the ocean floor and more

Check used bookstores for National Geographic maps that have become separated from the issues in which they originally appeared. Many second-hand bookstores put these in large, unsorted boxes and price them at less than $1 each. The physiographic maps of the ocean floor are real gems. Another to look for is one called *Earth's Dynamic Crust* (August 1985), which has a global tectonics map on one side and a dense summary of the tectonic evolution of the western US on the other. For a complete list of maps, write: National Geographic Society, Washington, DC 20036

T. Simkin, Tilling, R., Taggart, J. Jones, W. and Spall, H., 1989. *This Dynamic Planet : A world map of volcanoes, earthquakes and plate tectonics.* Copublished by the US Geological Survey and the Smithsonian Institution.

An excellent visual summary of all things tectonic. Highly recommended.

Sageware Corp, 1991. *TIME MACHINE EARTH*

An easy-to-use PC-compatible program that displays plate reconstructions for user-specified times in the geologic past. Current version is keyboard- rather than mouse-, driven and does not operate in the Windows environment. Approximately $180. Sageware Corp., 1282 Garner Ave., Schenectady, NY 12309. Telephone: 518/377-1052.

Volcanism:

Simarski, L., 1992. *Volcanism and climate change: An AGU Special Report.* Washington DC: American Geophysical Union, 27 pp.

A densely but clearly written summary of the current understanding of how volcanic emissions influence the global atmosphere and climate, with special emphasis on the 1991 eruption of Mt. Pinatubo. Also includes a brief discussion of how meteorite impacts might alter climate. A bargain at $4.00. ISBN 87590-818-7.

Earthquakes:

US Geological Survey, 1991. Safety and survival in an earthquake.

One of many public-interest pamphlets available from: US Geological Survey Books & Open File Section, Box 25425, Federal Center, Denver, CO 80225. Pamphlet # 1991-0-305-628.

Tectonics and human society:

King, G., Bailey, G., and Sturdy, D., 1994. Active tectonics, topography and human survival tactics. *Journal of Geophysical Research*, v. 99.

A provocative essay arguing that tectonics has been a major factor in shaping human cultures.

EARTH'S SURFACE IN FLUX

How do landscapes change through time?
How can ancient landscapes be reconstructed?

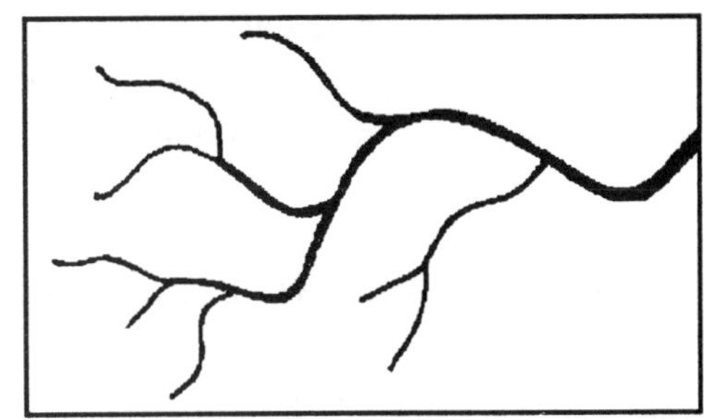

At the surface of the Earth, the tectonic system, driven by the planet's internal heat energy, meets the atmospheric system, driven by solar energy. Even as mountains rise at plate boundaries, they are worn down by the inexorable erosive power of water, wind and ice.

Everywhere on the planet, surficial processes continuously reorganize Earth materials and sculpt new landscapes. Water attacks rocks both physically and chemically, reducing them to progressively smaller pieces and carrying away their components in solution. These rock by-products may then be dispersed far from their origins. Over time, such processes have dismantled great mountain belts, carried their fragments to the continental shelves, and made the sea salty.

Although Earth's surface is always changing, vanished landscapes survive in sediments and sedimentary rocks. By observing sedimentary processes active today and making the Uniformitarian assumption that the same processes occurred in the past, it is possible to "read" the history of Earth's surface from ancient sedimentary deposits.

<table>
<tr><td><u>OBJECTIVES</u></td><td>

- To explore how running water shapes modern landscapes
- To consider the interplay between surficial and tectonic processes
- To discover how ancient land- and seascapes can be recreated through analysis of sedimentary deposits
- To investigate how the physical and chemical properties of sediments affect their practical uses

</td></tr>
<tr><td><u>OUTLINE</u></td><td>

I. Rivers as sculptors of landscapes
    Recurrent patterns at many scales
    The erosive power of rivers
    Ice and water

II. Sediments as records of ancient landscapes
    Clastic sedimentary rocks
    Chemical sedimentary rocks
    Sediments and sedimentary rocks as building materials

</td></tr>
</table>

LAB PREVIEW   Before coming to lab,
- Answer as many of the following questions as you can
- Read Chap. 9 ("Water on Land") and Chap. 11 ("The Changing Face of the Land) in *The Blue Planet,* or other readings assigned by your instructor

How much do you know already about processes that reshape the surface of the Earth?

What is a *fractal*? In what sense are rivers examples of fractals?

What is a continental divide?

How much of the continental U.S. is drained by streams and rivers of the Mississippi system?

What causes rivers to cut into rock and form canyons?

How do sandstone and limestone form?

What are glass, brick, plaster and concrete made from?

What types of sediment are the best choices for landfill sites?

EXERCISES

tributary:
river or stream
that flows into a
larger river,
contributing water
and sediment to it

*The courses of the
main Mississippi
tributaries, the
Ohio and Missouri
rivers, nearly
coincide with the
southern limit of
glaciation in the
last Ice Age. Why
do you think this
relationship exists?*

topographic map:
map that quantita-
tively illustrates
the relief of the
landscape (see
Appendix III)

fractal:
geometric feature
that looks the same
at a range of
scales

dendritic:
branching or tree-
like in form

I. Rivers as sculptors of landscapes

Except in very arid or very cold regions, running water is the principal shaper of continental landscapes. As rainwater seeks its way to the sea, it carves paths for itself, creating intricate networks of streams and rivers. Though riverscapes range from minor gullies to the Grand Canyon, rivers of all sizes share many basic characteristics.

A. Recurrent patterns at many scales

1. Study the physiographic map of North America and identify the Mississippi River and its principal <u>tributaries</u>. Sketch the map pattern of these major rivers:

Now examine a more detailed <u>topographic</u> map of your region. Identify the largest river in the area and its principal <u>tributaries</u>. Sketch the map pattern of these regional rivers:

Are the map patterns of the large and small river systems similar?

Describe their form.

2. River systems are good examples of <u>fractals</u> -- features that have the same geometry at many different spatial scales. Describe and/or sketch at least one other example of a fractal feature in the natural world that has a branching or <u>dendritic</u> form at many different scales. What do you think is the purpose of this branching design? In what sense is this similar to the function of a river system?

The word 'fractal', coined by Benoit Mandelbrot of IBM, is derived from the words 'fractional dimension'...

...A fractal feature like the map view of a river system is more than a 1-dimensional line but less than a 2-dimensional plane .

TABLE 4-1.
APPARENT
LENGTH OF THE

_ _ _ _ _ _ _ _
RIVER AT
DECREASING
RULER LENGTHS

3. An essential characteristic of fractals is that their size cannot be uniquely defined -- it depends on the scale at which the object is being measured. For example, the apparent length of a branching river system increases as the length of the "ruler" used to measure it decreases. This is because rivers, like all fractal features, show more and more detail the closer one looks.

Because they tend to have meandering, rather than straight, courses, the apparent lengths of individual rivers also tend to increase as the "ruler" used to measure them gets smaller. Choose a moderate-size river in your area and find out how its apparent length changes with ruler size, following the steps below.

a) Open a drafting caliper or compass to a setting close to its maximum and determine what distance this represents on the topographic map of your area. Enter this and the name of the river you are analyzing in Table 4-1 below.

b) Use the caliper or compass to 'walk' the length of the river you have chosen from one side of the map to the other. (At this 'ruler' length, you won't be able to follow the bends in the river very closely). Keep track of how many 'steps' you take (the last one will probably be a fraction of a step) and enter this in Table 4-1.

c) Close the caliper or compass by about a centimeter and repeat steps a and b. Then repeat the procedure until you reach the smallest possible setting.

Caliper setting (cm or inches)	Equivalent map distance (km or miles)	Number of "steps"	Apparent length of river (km or miles)

To convert the caliper setting to an actual distance, multiply by the map's scaling factor. For example, if the setting is 10 cm and the map scale is 1 : 25,000 then the actual distance is 10 x 25,000 cm = 250,000 cm = 2.5 km.

d) Did the apparent length of your river increase as the ruler size decreased? Make a plot of the apparent river length (y-axis) vs. ruler length (second column in Table 4-1, x-axis).

Would the apparent length increase with decreasing ruler length if the river were perfectly straight? Explain.

e) (Optional) The fractal dimension of your river can be determined by changing both axes on your plot to logarithmic scales and finding the slope (S) of the best-fit line. The fractal dimension (D) is defined as:

$$D = (1 - S).$$

(4-1)

The slope **S** should be negative and have an absolute value between 0 and 1, so the fractal dimension will be a number between 1 and 2.

drainage basin:
the land area
drained by a river
or river system

FIGURE 4-1.
DRAINAGE
BASINS ON AN
IMAGINARY
CONTINENT

divide:
topographic high
defining the
boundary of the
drainage basin

 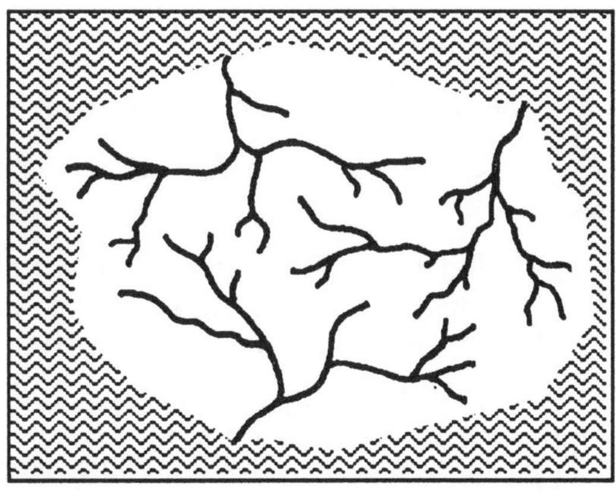

*For centuries, rivers
were the highways
for inland travelers.
Drainage divides
dictated patterns of
trade and cultural
exchange.*

4. A <u>drainage basin</u> is the land area drained by a particular river or river system. The boundaries of a drainage basin are topographic highs called <u>divides</u>. For a major river system like the Mississippi, these boundaries are called *continental divides*, since raindrops falling on opposite sides of the boundary will take completely different paths to the ocean.

a) Outline the continental divides on the imaginary land mass in Figure 4-1.

b) Using a transparent overlay on the topographic map of your region, outline the drainage basin of a local stream or river.

Imagine that a chemical spill occurs at some point along that river. Make an X to mark the location of such a spill and show which other streams would be contaminated by it. Which streams would be unaffected and why?

c) On the physiographic map of North America, trace some part of the boundary of the Mississippi drainage basin, whose total area is 3.18 million km^2 -- about 40% of the area of the continental US. What major topographic features form the eastern and western continental divides? What is surprising about the northern and southern boundaries?

Do you live within the Mississippi drainage basin?

If not, which major river basin do you live in? (i.e., which river ultimately delivers the water of local streams to the sea?)

Trace the route water takes from local creeks and rivers to the sea. List the names of these progressively larger rivers. How many times do small rivers join larger ones on their way to the sea?

gradient:
the slope of a
river channel,
expressed in feet
per mile, meters
per kilometer, or
as a percentage

gradient:
the slope of a
river channel,
expressed in feet
per mile, meters
per kilometer, or
as a percentage

river profile:
cross-sectional
view of a river
channel in the
downstream
direction

B. The erosive power of rivers

Rivers carry more than water to the sea; they also transport tremendous volumes of sediment. But unlike river water, which moves constantly downstream and reaches the sea in days or weeks, river sediments may be temporarily deposited and stay in the river system for many centuries before moving further downstream. The rate of sediment transport through river systems depends on several factors.

1. The gradient of a river -- the slope of its channel -- is a major factor governing its ability to shape the landscape. In general, as a river's gradient steepens, its water flows faster and has a greater capacity to pick up and carry sediments.

Gradient is the drop in elevation over some distance downstream:

$$\text{Gradient (m/km or ft/mile)} \quad = \quad \frac{\text{Drop in elevation (m or ft)}}{\text{Downstream distance (km or miles)}} \qquad (4\text{--}2)$$

a) Determine the average gradient of rivers from your area to the sea (or, if appropriate, to a large inland lake). First, find the elevation of your city on the local map. (If you are unfamiliar with topographic maps, see Appendix III).

Drop in elevation to sea (lake) = _________ ft = _________ m

Next, turn to the physiographic map of the continent and estimate the total downstream length of river segments from your location to the sea (lake). Use equation 4-3 and show your calculations.

$$\begin{array}{llll} \text{Total downstream} = & \text{Summed distance} & \text{x} & \text{Map scaling} & (4\text{-}3) \\ \text{distance} & \text{on map} & & \text{factor} \end{array}$$

Total downstream distance = ______________

Finally, determine the average gradient in m/km or ft/mile using equation 4-2.

Average gradient = ______________

b) In general, the gradient of a river system is not uniform over its entire length. Using the local topographic map and the procedure above, determine the gradient of a segment of a river in your area. If you live very close to the sea (or large inland lake), find the gradient of a stream or river some distance upstream.

Drop in elevation over stream segment   =   __________
Downstream distance along segment   =   __________
Local stream gradient   =   __________

How does the local stream gradient compare with the average gradient of the system downstream from your location?

c) Draw a schematic downstream profile of the river system illustrating how the channel gradient changes with distance downstream:

Elevation

-------------Distance downstream --------------->

2. Consider the implications of the shape of the river profile you have drawn.

a) Given the downstream trend in channel gradient, where do you think river systems have their greatest power to erode and carry sediments? Where will they tend to deposit sediments?

b) How will this change the shape of a river profile through time? Illustrate the change with a dotted line on the profile you drew in exercise B.1.c.

c) Over time, these patterns of erosion and deposition should have leveled the topography of the continents, leaving them flat and featureless. What processes prevent this from happening?

delta:
sediment deposited at the mouth of a river as it enters a standing body of water. Often triangular in shape, resembling the Greek letter delta.

3. A <u>delta</u> forms where a river enters the sea or other standing body of water. The Mississippi delivers tens of thousands of tons of sediment to the Gulf of Mexico *each day*, and the delta grows seaward at a rate of 2 km/decade. The position of the delta has changed repeatedly, however, in the last few centuries; most of southern Louisiana is made of deltaic sediments.

a) What do you think causes parts of the delta to be abandoned? (Consider what happens to the river's gradient as the delta extends seaward).

At times in the geologic past, the Atchafalaya did carry much of the Mississippi's water to the Gulf, and its sediments built the bayou country of SW Louisiana.

b) The gradient of the lower Mississippi is getting so low that the river is seeking a shorter, steeper route to the sea. The most likely place for this to happen is at the confluence, or joining, of the Red River with the Mississippi east of Marksville, Louisiana. Left to itself, Mississippi water would flow northwest along a short segment of the Red River then into the Atchafalaya river and to the Gulf. Because this would be disastrous for New Orleans, the US Army Corps of Engineers spends millions of dollars each year on an elaborate series of dams and water gates that keep the Mississippi in its present configuration.

The elevation at the confluence of the Red and Mississippi Rivers is only 5 m above sea level. Use the physiographic map of the US to determine the gradient from this point to the sea along 1) the Mississippi and 2) the Atchafalaya.

Most of the world's coal originated as woody material buried in deltaic sediments.

	Mississippi	Atchafalaya
Drop in elevation from Red R. to sea	5 m	5 m
Downstream distance from Red R. to sea		
Gradient from Red R. to sea		

sediment fan:
fan-shaped accumu-
lation of river
sediments at the
base of a steep
slope. If at the
base of the contin-
ental shelf, termed
a *submarine fan*;
at the foot of a
mountain on land,
an *alluvial fan*.

4. The Gulf of Mexico is the ultimate repository for sediments collected by the Mississippi River and its tributaries. These organic-rich sediments are the source of much of the domestically-produced oil in the US. It is estimated that the total volume of sediment that has been delivered by the Mississippi to the Gulf is more than 3×10^5 km^3.

On the map of the ocean floor, find the submarine <u>sediment fan</u> produced by the Mississippi, then identify fans associated with other major rivers around the world. List 5 or 6 of the largest fans and the rivers that formed them. Why do you think the fans in the Arabian Sea and Bay of Bengal are especially large?

C. Ice and water
During ice ages, sea level falls significantly because much of the planet's water is locked in glaciers and ice caps. Even in areas that remain unglaciated, this has a profound effect on river behavior.

base level:
elevation of the
body of water into
which a river flows

1. Rivers can erode no deeper than the elevation of the body of water they enter -- their <u>base level</u>. The ultimate base level for all rivers is, of course, sea level. How would you expect rivers to respond to the drop in sea level that occurs during an ice age? Explain your reasoning and illustrate your answer by drawing a sketch showing how the profile of the river looks: a) shortly after sea level has fallen (dotted line) and b) sometime later (solid line).

River profiles:

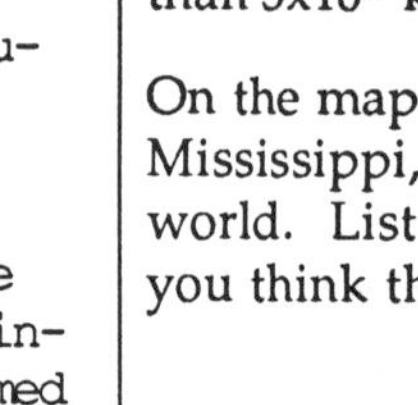

Old sea level

New sea level

-------------Distance downstream -------------->

2. At the end of the last ice age, sea level rose dramatically as the great ice caps melted (you'll calculate the actual rise in Laboratory 5). River systems on the east coast of the US were 'drowned', creating <u>estuaries</u> like Chesapeake Bay.

a) Find Chesapeake Bay on the physiographic map of the US and sketch its outline. What aspect of its shape suggests that it is a submerged river channel?

estuary:
bay formed when
rising sea level
submerges the lower
reaches of a river

b) Estuaries are not characteristic of the west coast of the US. How can you explain this? Consider the tectonic settings of the east and west coasts.

II. Sediments as records of ancient landscapes

As you discovered in Laboratory 2 (*Earth in Time*), sediments and sedimentary rocks constitute important archives of the geologic past. Not only are they the foundation for the fossil-based geologic time scale, but they also provide information about what the surface of the Earth was like at times in the past.

A. Clastic sedimentary rocks

<u>Clastic</u> sedimentary rocks were deposited physically -- i.e. as discrete particles -- by water, wind or ice. Many have physical features that provide clues to the environments in which they were deposited.

1. Examine the specimens in the lab. Although these rocks are millions of years old, similar structures are formed in sediments being deposited today. Describe and sketch each specimen, then use a uniformitarian point of view to speculate about where and how the rock formed. As you study the rocks, consider the following questions:

> Which specimens appear to have been laid down by moving water?
> How might you determine the relative speed of the water?
> Which specimen(s) must have been deposited on land?
> Which show evidence for the presence of organisms?
> Which have features that indicate which direction was
>   'up' at the time the sediments were deposited?

---

clastic sediment:
sediment deposited physically by water, wind or ice

CLASTIC SEDIMENTARY ROCKS:

conglomerate:
rock composed of pebbles and larger grains (>2 mm); deposited by fast-flowing water

sandstone:
rock composed of sand-sized grains (1/16-2 mm)

siltstone:
rock composed of silt-sized grains (1/16 -1/256 mm)

shale/mudstone:
rock composed of clay-sized grains (<1/256 mm); deposited in quiet bodies of water

diamictite:
rock composed of two distinct sizes of clastic grains; may represent a glacial deposit or mud flow

---

SEDIMENTARY STRUCTURES:

bedding:
depositional layering

ripple marks:
ripples formed in sand by moving water or air

desiccation cracks (mudcracks):
polygonal cracks formed on the surface of fine-grained sediment as it dries and contracts

cross bedding:
layering at an angle to overall bedding in a rock, representing the downstream faces of ripples or dunes

2. Grain characteristics and depositional setting
Like sedimentary structures, the sizes and shapes of grains in clastic sediments are also important clues to the environment in which they accumulated.

a) Think about how you would expect grain sizes and shapes to vary in sediments deposited in different ways.

Which would probably contain coarser (larger) grains: Sediments deposited in a quiet lake or sediments deposited on a wave-washed beach? Why?

well-sorted sediment:
sediment in which grain size is uniform

Which would be more uniform in grain size (<u>well-sorted</u>): Sediments deposited in a flowing river or sediments deposited directly by glacial ice? Why?

In which would grains tend to be more rounded (lacking sharp edges and corners): A conglomerate whose pebbles came from source rocks close to the site of deposition or one whose pebbles traveled far before deposition? Why?

b) Examine small amounts of the sediment samples with a hand lens or microscope. Describe the size, sorting and shapes of the grains. If sieves and balances are available, use them to make quantitative estimates of grain size distribution.

Record your observations in the table below. From the list of depositional settings provided by your instructor, identify the most likely setting of origin for each sediment sample. Leave the "permeability" column empty for now.

TABLE 4-2.
GRAIN SIZE AND SHAPE CHARACTERISTICS FOR SEDIMENT SAMPLES

Sample	Grain size range	Sorting (good/poor)	Grain shape (angular/rounded)	Depositional setting	Permeability (low/med/high)
1					
2					
3					

porosity:
amount (percent)
of pore space in a
sediment or rock

permeability:
measure of the ease
with which fluids
can move through
sediment or rock

groundwater:
water contained in
the pore spaces of
soil, sediment and
rock

3. Permeability

Grain size, sorting and shape control bulk sediment properties that are important to humans who build on, draw water from, and dispose of wastes in sediments. Two such properties are <u>porosity</u> and <u>permeability</u>. Porosity is the total amount of pore space or empty space in a sediment or rock. Permeability is a measure of the ease with which fluid (for example, water or oil) can move through sediments and rocks. Permeability has to do with the interconnectedness of pore spaces in the material. A highly porous rock could be relatively impermeable if its pore spaces were isolated from each other.

Knowledge of permeability is critical in drilling productive wells; even if water or oil is abundant in the subsurface, it will not flow if permeability is low. Permeability is also an important consideration in finding suitable waste disposal sites. Waste must be placed in low-permeability geologic materials if it is not to contaminate local rivers, lakes or <u>groundwater</u>.

In this exercise, you will explore the relationship between permeability and grain size in clastic sediments by making qualitative permeability determinations for the sediment samples you examined above.

a) Think about how you would expect permeability to vary with grain size. Which do you think would be more permeable: Coarse or fine sediment? Well-sorted or poorly sorted sediment? What other factors might affect permeability? Explain your logic and illustrate your answer if appropriate.

b) Test your predictions by recording how long it takes for water to pass through a given volume of each of the sediment samples, following the procedure demonstrated by your instructor. Enter the relative permeability values in the table on the previous page. Are your results what you expected? If not, revise your statements about how grain size, sorting and shape affect permeability.

c) Which sediment type would be most likely to act as a subsurface <u>aquifer</u>, or layer through which groundwater flows? Why? What is the nature of the aquifers in your area?

aquifer:
permeable layer of
sediment or rock
that carries
groundwater

Which sediment type would be the best choice for containing hazardous waste? Explain.

chemical sediment
sediment formed by chemical precipitation of minerals dissolved in water

limestone:
rock composed mainly of calcite ($CaCO_3$); usually precipitated from seawater

travertine:
rock composed of calcite precipitated from water at hot springs or in caves

evaporite:
rock composed of salts including halite, gypsum and anhydrite, formed by evaporation of saltwater

chert, agate:
rocks formed by precipitation of dissolved silica (SiO_2) from sea-water, freshwater or groundwater

The manufacture of Portland cement from limestone drives carbon dioxide (CO_2) out of the calcite, and is one source of human contributions to atmospheric CO_2 (but is minor compared with that produced by burning of fossils fuels).

B. Chemical sedimentary rocks

<u>Chemical</u> sedimentary rocks were deposited by chemical precipitation of minerals dissolved in water. Like clastic sedimentary rocks, their characteristics provide clues to the environments in which they accumulated.

Examine, then describe or sketch, the specimens of chemical sedimentary rocks. Which represent marine deposits? Which may have accumulated in land environments?

C. Sediments and sedimentary rocks as building materials

Since ancient times, sediments and sedimentary rocks have been used as building materials. The room in which you are sitting probably consists largely of products derived from sediments. Examine the manufactured materials and match them with the natural sedimentary materials from which they were made.

Glass is made by fusing (melting) and then cooling silica (quartz) sand from unconsolidated deposits or very pure sandstones.
Brick and tile consist of clay hardened by high-temperature baking. Clay is derived from settings including lake beds, estuaries and marine environments.
Plaster is made by heating gypsum ($Ca\,SO_4 \cdot H_2O$) to dehydrate it, then crushing it into a fine powder.
Concrete typically consists of sand and pebbles set in a cement made from limestone and clay that have been fired in a kiln.

Are any of these products manufactured in your area? If so, where do the raw materials come from? What were their original depositional settings?

III. Synthesis

A. Preview review
Return to the Lab Preview and complete any questions you were unable to answer before.

B. Rivers in art and literature
The constant motion and change that is inherent to rivers makes them seem almost animate. The personalities of particular rivers have inspired authors, artists and composers to create their portraits.

Choose an artistic work (book excerpt, poem, painting, song...) whose subject is a river, and describe how the work captures some aspect of the river's character. Here are examples:

> <u>Books</u>
> *Life on the Mississippi* by Mark Twain, Ch. 27: "Cutoffs & Stephen"
> *The Control of Nature* by John McPhee, Part I: "Atchafalaya" (see references below)
> <u>Music</u>
> *The Moldau* by Czech composer Bedrich Smetana
> <u>Paintings</u>
> *The Mill*, Rembrandt
> *The Oxbow*, Thomas Cole
> *Canoe in the Rapids*, Winslow Homer

C. River watch
Write a short report based on one of the following suggestions:

- Observe a river in your area and take note of processes of sediment transport and deposition. Where are sediments picked up by the water? Where are they dropped? Is the river level high or low compared to its usual level? Can you see evidence of recent high-water events? Collect a sample of river water. Is there any sediment in it?

- Find information on the history of a river in your area. How often has it flooded in the past century? When were any dams or locks built and how did these affect the river's behavior both up- and downstream? A local historical society may be a good source of information.

- Focus on some aspect of the Mississippi Valley floods in the summer of 1993 -- e.g. estimates of property damage and crop loss or the effect the flooding had on a particular town. Refer to newspaper and magazine articles from June, July and August 1993 for information. If you experienced the flooding yourself, write about how it affected your day-to-day life.

D. Other worlds
Would you expect to find sedimentary rocks on the Moon, Venus or Mars? Review the discussion of these bodies in Laboratory 1 and explain your reasoning.

SUGGESTED REFERENCES AND RESOURCES

Bass, R., 1989. *Oil Notes*. Boston: Houghton-Mifflin. 172 pp.
> The personal journal of a petroleum exploration geologist, with lyrical passages about the pleasures of "seeing" ancient landscapes in the sediments of a drill core.

Dietrich, R. and Skinner, B., 1990. *Gems, Granites and Gravels: Knowing and using rocks and minerals*. Cambridge: Cambridge University Press. 173 pp.
> An introductory geology text that emphasizes the origins of geological materials with practical uses. The chapters on "Soils, dust and mud" and "Building materials" are particularly appropriate to the material covered in this laboratory.

Hsü, K. 1983 *The Mediterranean Was a Desert*. Princeton Univ. Press. 197 pp.
> An engaging account of the *Glomar Challenger* drilling cruise that discovered thick evaporite deposits in the Mediterranean basin. An excellent example of how sedimentary deposits are the key to reconstructing ancient land- and seascapes.

McPhee, J., 1989. *The Control of Nature*, Part I: Atchafalaya. New York: Farrar, Strauss and Giroux.
> The saga of the Army Corps of Engineers' "war" with the Mississippi at the point where it threatens to defect to the channel of the Atchafalaya. Written with McPhee's characteristic perceptiveness about the peculiarities of both nature and human nature.

Erwin Raisz Landform Maps of the US, Europe, Asia and Africa.
> These classic black and white physiographic maps are rich in detail and, when laminated, can be written on with transparency pens and used in a wide variety of ways. For example, students can outline the boundaries of major drainage basins or follow the path of local rainwater from area creeks and rivers to the sea. The maps come in a wide range of dimensions, from page size to about 2'x 3'. Distributed by Raisz Landform Maps, P.O. Box 773, Melrose, MA 02176. Phone: 800/242-3199; FAX: 617/662-2622.

Sheriff, W., and Rasband, W., 1993, *ImageFractal* National Institutes of Health. Macintosh software in the public domain.
> *ImageFractal*, a fractal analysis program, is an extension of the versatile *Image* software originally developed at NIH for quantitative geometric analysis of biological images. Both programs have tremendous potential for geoscience applications and are available over the Internet through the NIH Gopher. Images to be analyzed must be 8-bit TIFF files.

Equipment and supplies for sediment analysis:
> <u>Portable sediment sieves:</u> A small, lightweight and inexpensive ($50) set of 5 stacking sediment sieves is available from Keck Instruments, Inc., a supplier for environmental consulting firms. Product # SS-81 (Keck Sand Shaker). Address: 1099 W. Grand River Ave., P.O. Box 345, Williamston, MI 48895. Phone: 800/542-5681; FAX: 517/655-1157.
>
> <u>Grain characterization folder:</u> A pocket-sized reference designed to help the user make quick, accurate descriptions of grain size, shape and sorting. Available from Forestry Suppliers, Inc. Product #77332, approximately $6. Address: 205 W. Rankin St., P.O. Box 8397, Jackson, MS 39284. Phone: 800/647-5368; FAX: 800/543-4203.

EARTH'S AURA: THE ATMOSPHERE AND HYDROSPHERE

How are Earth's air and water linked with its interior?
Are human-induced changes in the atmosphere and
*   hydrosphere significant relative to natural variations?*

Viewed from space, Earth is a blue planet, a jewel-like sphere that might more appropriately be called "Ocean" than "Earth", since so much of its surface is covered by water. In Laboratory 1, you discovered significant ways in which Earth is different from the other planets of the Solar System. Among the inner (terrestrial) planets, only Earth, with its peculiar atmospheric composition and surface conditions, has abundant liquid water.

Another important distinction is that the Earth's air and water are not static, but in a constant state of flux. Elements are cycled and recycled through the atmosphere, hydrosphere, biosphere and solid earth. In this laboratory, you will explore two of these interrelated chemical cycles, the hydrologic (water) and carbon cycles, on global and local scales.

<u>**OBJECTIVES**</u>
- To explore the Earth's water and carbon budgets in terms of reservoirs, fluxes and residence times
- To discover connections between atmospheric, hydrologic, and tectonic processes
- To evaluate the role of human activities in climate change

<u>**OUTLINE**</u>
I. The hydrologic (water) cycle
   Global reservoirs and fluxes
   Local hydrologic budgets
   Water and tectonics

II. The carbon cycle
   Global reservoirs and fluxes
   Local carbon fluxes
   Climate and the carbon cycle
   Human effects on climate

<u>LAB PREVIEW</u>   Before coming to lab,
- Answer as many of the following questions as you can
- Read the following chapters in *The Blue Planet*:
    - Chapter 8:   "The World Ocean"
    - Chapter 12:   "Composition and Structure of the Atmosphere"
    - Chapter 14:   "The Earth's Changing Climates"
    - Chapter 18:   "Global Change: A Planet Under Stress"

Or other readings assigned by your instructor

How much do you know already about Earth's air and water?

What percentage of Earth's water is in the oceans?

Which contain more of Earth's water:
  a) Ice caps and glaciers or b) All lakes and rivers combined?

How long does an average drop of water stay in the atmosphere?

In the subsurface as groundwater?

In the oceans?

What is the greenhouse effect?

What is meant by positive feedback? Negative feedback?

What evidence indicates that Earth's climate has changed in the past?

What are some possible causes of these natural climatic changes?

What human activities have the potential to have greatest impact on Earth's atmosphere and climate? What consequences could follow if these activities are not curbed?

EXERCISES

hydrologic cycle:
the continuous cycling of water in different forms at, above and under Earth's surface

reservoir:
temporary "storage place" for a particular material

FIGURE 5-1.
PRINCIPAL COMPONENTS OF THE HYDROLOGIC CYCLE ON EARTH
(See also Tables 5-1 and 5-2).

flux:
process that moves a material from one reservoir to another at a particular rate

I. The hydrologic cycle

The water you washed with this morning is old and well traveled. At times in the past, it has resided in oceans, clouds, glaciers, rivers and lakes, rocks and soil, plants and animals. The water now at and near the surface of the Earth is the same that fell as rain, flowed in rivers, and filled lakes and seas on a young Earth four billion years ago.

The cyclic movement of water on, in and above Earth is termed the <u>hydrologic cycle</u>. The cycle can best be understood by identifying <u>reservoirs</u>, (places where water is temporarily held), and <u>fluxes</u>, (processes that move water from one reservoir to another). These are summarized in the figure below and Tables 5-1 and 5-2 on the next page.

A. Global water reservoirs and fluxes

1. Complete the sketch of the water cycle below by drawing arrows to represent the fluxes in Table 5-2 . Draw heavier arrows to show the processes that move the largest amounts of water each year. Label the arrows with descriptions of the processes they represent (e.g. precipitation, transpiration).

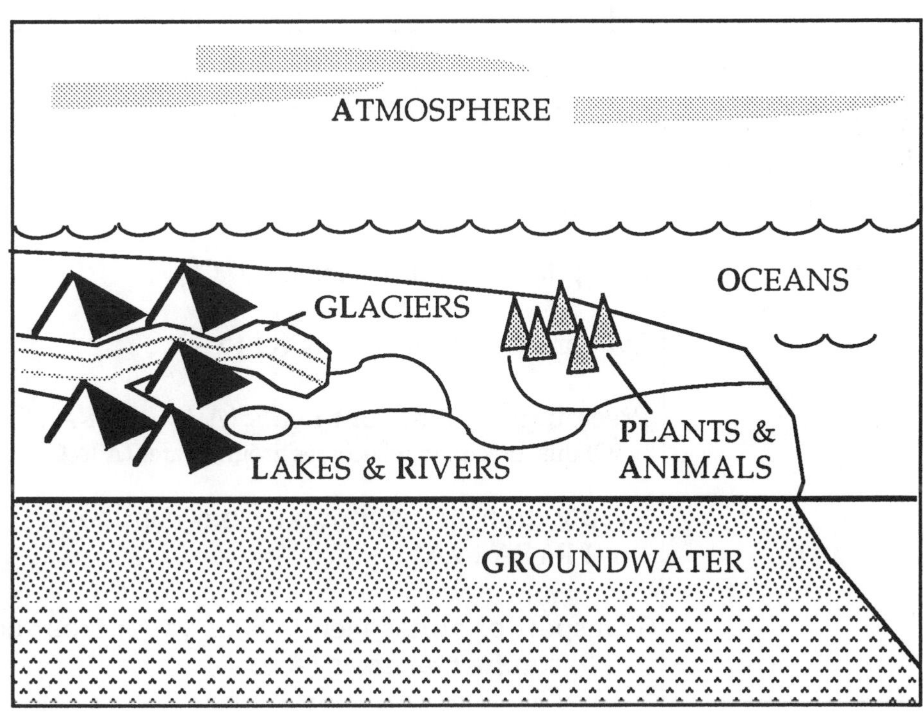

2. Would you expect there to be a comparable hydrologic cycle on Venus? Explain.

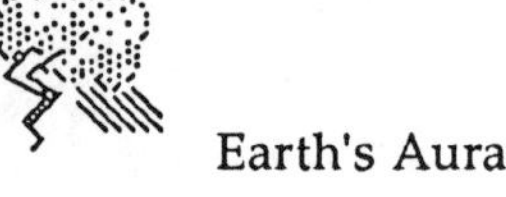

TABLE 5-1.
WATER RESERVOIRS ON EARTH

Reservoir	_Volume ,(km^3)_	_% of Earth's water_
Oceans (**O**)	1,320,000,000	97.21
Glaciers (**Gl**)	29,200,000	2.15
Groundwater (**Gr**) and soil moisture	8,417,000	0.62
Lakes & rivers (**LR**)	230,000	0.017
Atmosphere (**A**)	13,000	0.001
Plants & animals (**PA**)	<10,000	<0.00075
TOTAL	1,357,870,000	100

TABLE 5-2.
WATER FLUXES ON EARTH

Description of process	_Volume of water moved (km^3/year)_
Precipitation into oceans (**A to O**)	385,000
Evaporation from oceans (**O to A**)	425,000
Precipitation onto continents (**A to LR+PA+Gr+Gl**)	111,000
Of this, precip. that becomes groundwater (**A to Gr**)	(16,000)
Evaporation, transpiration from continents (**LR+PA to A**)	71,000
Transport of atmospheric water vapor from oceanic to continental areas (**A to A**)	40,000
Flow of surface, groundwater to oceans (**LR+Gr to O**)	40,000
Of this, groundwater discharge to oceans (**Gr to O**)	(16,000)
TOTAL ANNUAL FLUX	1,072,000

Sources: Skinner, B. and Porter, S., 1987. _The Dynamic Earth._ New York: Wiley.
Maurits la Riviere, J., 1989. Threats to the world's water. _Sci. Am._ v. 261, no. 3, p. 80-94.

3. What fraction of Earth's water is in motion in a given year? To calculate this, divide the total annual water fluxes by the total volume of the water reservoirs.

4. According to Table 5-2, how much water enters the oceans annually?

How much water leaves the oceans annually?

Are the inflow to and outflow from the oceans equal? What would it mean if they were not -- for example if inflow were greater than outflow?

Check to see if equal amounts of water enter and leave the atmosphere each year. Show your calculations.

residence time: average length of time a material remains in a particular reservoir

Ocean water contains enough dissolved carbon dioxide that it can be dated by the C-14 method. Deep ocean waters give ages of more than 5000 yrs -- the length of time since the water last circulated through the atmosphere.

5. By comparing the size of a given reservoir with the rate at which water enters or leaves it, we can get a sense of the <u>residence time</u> of a typical drop of water in the reservoir. That is:

$$\text{Residence time} = \text{Reservoir size} / \text{Total inflow (or outflow)} \qquad (5\text{-}1)$$

What is the average residence time of ocean water?

Of groundwater?

Of water in the atmosphere?

6. During the last Ice Age, the volume of (liquid-equivalent) water in glaciers and ice caps was 4.38×10^7 km^3 greater than today. What do you think happened to this water when the great glaciers melted -- i.e. what reservoir did it enter?

7. Assuming that the area of the oceans is 3.5×10^8 km^2 and that the ocean basins have approximately vertical sides (Fig. 5-2), estimate how much sea level rose at the end of the last Ice Age. Show your calculations.

FIGURE 5-2.
SEA LEVEL RISE
AT THE END OF
THE ICE AGE

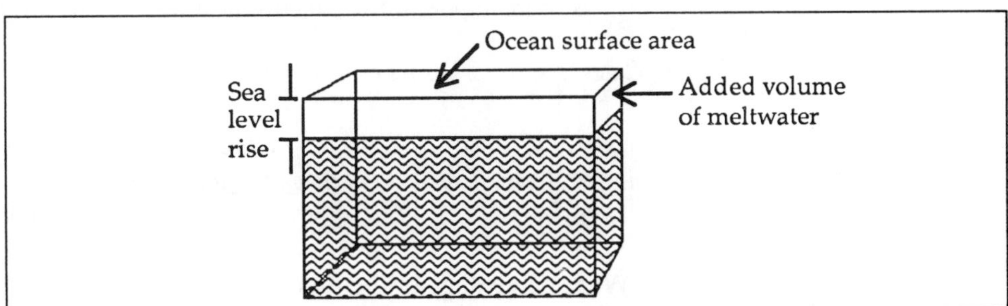

**hydrologic bud-
get equation:**
equation that
accounts for all
water entering or
leaving an area

gauging station:
station on a river
where water height
and flow rate are
monitored

B. Local hydrologic budgets

On a local scale, the hydrologic cycle can be described with an input-output equation called the <u>hydrologic budget equation</u>. This equation accounts for all the water that enters and leaves a given area. Over a period of decades, there should be a balance between the incoming and outgoing water fluxes for the region.

1. Write a word equation that balances the water budget for your region. Consider which parts of the global water cycle (Fig. 5-1) are important locally.

$$\text{INPUT} \quad = \quad \text{OUTPUT}$$

2. [**To be completed only if data are available. Answer on a separate sheet.**]
To apply the hydrologic budget equation, it is necessary to define the area in which water fluxes will be measured. A logical area to choose is a drainage basin -- the area that contributes water to a single stream or stream system (see Lab 4). In an ideal drainage basin, all water that falls as precipitation and does not evaporate into the air or percolate into the ground will leave the area at 1 well-defined point where it can be accurately measured at a stream <u>gauging station</u>.

• On the local topographic map, trace the boundaries of the basin to be analyzed. What is the name of the stream whose flow is monitored at the gauging station?

• What is the area of the basin? (Information provided by your instructor)
   Basin area (in m^2) = _________   *(= Area in km^2 x 10^6 m^2/km^2)*

• Use monthly weather service records and the following equation to determine the total volume of precipitation received in the basin in one calendar year. (For spreadsheet instructions, see Appendix I, p. 173).

$$\frac{\text{Volume of}}{\text{precip. } (m^3)} = \frac{\text{Total precipitation}}{\text{(sum of monthly values, in cm)}} \quad x \quad \frac{\text{Area of basin}}{(m^2)} \quad x \quad \frac{1\ m}{100\ cm} \quad (5\text{-}2)$$

   What assumptions make this only a rough estimate of the total precipitation?

• Use monthly evaporation data to estimate the total amount of water lost to the atmosphere (usually determined by recording water lost from a standing pan).

$$\frac{\text{Volume of water}}{\text{evaporated } (m^3)} = \frac{\text{Total evaporation}}{\text{(sum of monthly values, in cm)}} \quad x \quad \frac{\text{Area of basin}}{(m^2)} \quad x \quad \frac{1\ m}{100\ cm} \quad (5\text{-}3)$$

   What percentage of the precipitation left the basin by evaporation?

• Calculate the volume of water that left the basin as stream flow:

$$\frac{\text{Volume of water}}{\text{leaving as stream flow}} = \frac{\text{Sum of monthly}}{\underline{\text{flow rate values } (m^3/sec)}} \quad x \quad 3.16 \times 10^7\ sec/yr$$
$$(m^3/yr) \qquad 12\ \text{(to obtain average flow rate)} \qquad (5\text{-}4)$$

   What percentage of the precipitation left the basin by stream flow?
   What percentage is still unaccounted for? Where might this water have gone?

C. Water and tectonics

Water plays fundamental roles in many of the tectonic processes of the solid Earth. Seismicity and volcanism on Earth would be very different in the absence of liquid water. For example, the presence of groundwater in rocks greatly reduces their strength, making them more likely to fail brittlely. (Conversely, the absence of water on Venus may be one reason that there are no well-defined plate boundaries on the planet). At mid-ocean ridges, seawater circulates through new ocean crust, altering the newly formed rocks and picking up nutrients that sustain communities of exotic deep-sea organisms. In subduction zones, the water carried into the mantle by the downgoing oceanic plate significantly changes the melting temperature of mantle rock and gives rise to the explosive volcanic activity at convergent plate boundaries.

1. Role of water in subduction-related volcanism

Figure 5-3 is a <u>phase diagram</u> illustrating the pressure and temperature conditions under which rock will melt. Note that temperature increases to the right on the horizontal axis and pressure increases downward on the vertical axis, just as it does with increasing depth in Earth's interior. The straight line labeled A is the melting curve for rock in the absence of water. To the left of this line (at relatively low temperatures), rock is solid. To the right, rock melts.

phase diagram:

plot showing the physical conditions under which the different states of a substance (solid, liquid, or gas forms) exist

FIGURE 5-3.
SIMPLIFIED PHASE DIAGRAM FOR ROCK IN EARTH'S UPPER MANTLE

Water carried into the mantle by subducted oceanic plates includes water both physically trapped in sediments and chemically bound in clay minerals.

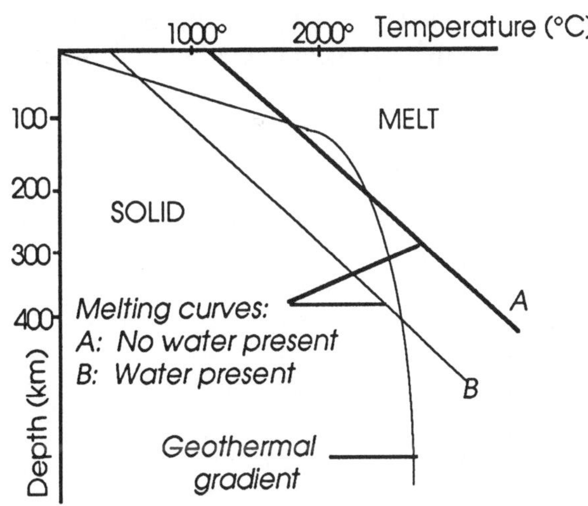

a) Describe how the melting temperature of rock varies with pressure (depth).

b) The line marked B shows the temperatures required for melting in the presence of water. What does the position of this line (relative to line A) indicate about the effect of water on the melting temperature of rock?

geothermal gradient: increase in temperature with depth in Earth's subsurface, expressed as °C/km

Water that is subducted with oceanic crust is one reason for the explosive nature of volcanism at convergent plate margins.

Ice skating is possible only because of the fact that the melting temperature of ice decreases with increasing pressure. The pressure of the blade creates a thin film of liquid water on which the skater glides.

c) The curved line in Figure 5-3 is the <u>geothermal gradient</u> -- the actual variation in temperature with depth in Earth's interior. In the absence of water, there is a limited depth range over which the geothermal gradient crosses the melting curve (A) into the melt field. A small amount of molten rock is present in the mantle at these depths.

What is the range of depths over which melting occurs in the absence of water? Indicate on Figure 5-3 how you determined this.

d) What is this narrow zone of partly molten mantle called?
What role does it play in isostasy and tectonics? (Review pp. 42-43 in Lab 3).

e) Over what depths will melting occur if water is present?
On Figure 5-3, label the depth range over which the geothermal gradient crosses into the melting field in the presence of water.

f) What is the tectonic significance of the area you labeled in question e?

2. Uncommon properties of a common substance
The phase diagram for rock (Figure 5-3) is similar to that for other substances -- that is, the melting temperature increases with pressure, indicating that the solid state is denser than the liquid (molten) state. Is this true for water? Cite an everyday observation that supports your answer.

Name one reason this unusual property is important in the natural world.

II. The carbon cycle

Like the hydrologic cycle, the carbon cycle is intimately linked with all Earth systems. The cycling of carbon through the atmosphere, hydrosphere, biosphere and solid earth is summarized in Figure 5-4 and Tables 5-3 and 5-4 (next page).

A. Global reservoirs and fluxes

1. The bulk chemical composition of Venus is similar to Earth's, so there should be equal amounts of carbon on the two planets. The distribution of that carbon is quite different, however. What carbon reservoirs exist on Earth but not on Venus? (See Fig. 5-4 and discussion of Venus in Lab 1).

What reservoir probably holds most of Venus' carbon?

2. In the absence of human activities, what fraction of Earth's carbon is in motion in a given year? See Tables 5-3, 5-4.

3. Compare the rates of creation (precipitation) and destruction (weathering) of carbonate rocks. Is the carbonate rock reservoir growing or shrinking? How fast?

What is the residence time of carbon in carbonate rocks (e.g. limestone)? Use the rate of carbonate weathering in your calculation.

4. Which carbon reservoir is rapidly being depleted by humans? By how many times do output fluxes exceed input fluxes?

FIGURE 5-4. PRINCIPAL COMPONENTS OF THE CARBON CYCLE

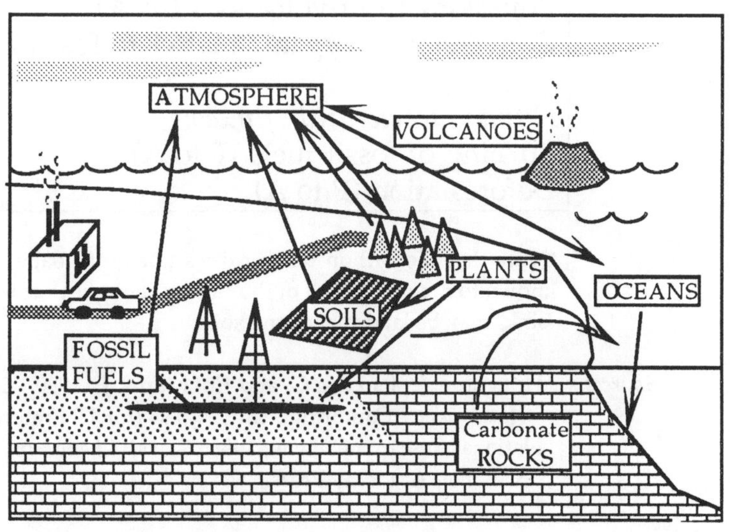

TABLE 5-3.
CARBON RESERVOIRS ON EARTH
1 metric ton= 1000 kg

Reservoirs	Mass in billion (10^9) metric tons
Carbonate rocks (**R**)	
Limestone ($CaCO_3$) and dolomite $(Ca,Mg)(CO_3)_2$	10,000,000
Oceans (**O**)	
As dissolved carbonate (CO_3) and bicarbonate (HCO_3)	36,000
Fossil fuels (**F**)	
Coal, oil and natural gas	6000
Soils (**S**)	
As organic matter and methane (CH_4)	1500
Atmosphere (**A**)	
Mainly as carbon dioxide (CO_2)	735
Plants (**P**)	560
On land and in the oceans	

TABLE 5-4.
CARBON FLUXES ON EARTH*

	Billions of metric tons/year
Natural fluxes:	
Photosynthesis by plants (**A** to **P**)	102
Respiration by plants (**P** to **A**)	50
Decomposition of plants (**P** to **S**)	50
Diffusion from soil to air (**S** to **A**)	50
Diffusion from atmosphere to ocean (**A** to **O**)	92
Diffusion from ocean to atmosphere (**O** to **A**)	90
Precipitation of carbonate rocks in ocean (**O** to **R**)	2.2
Weathering of carbonate rocks (**R** to **O**)	0.8
Metamorphism of carbonate rocks (**R** to **A**)	0.1
Weathering of silicate rocks (**A** to **R** to **O**)	0.1
Emissions from volcanoes (to **A**)	0.2
Conversion of organic matter to fossil fuels (**P** to **F**)	<1
Human-accelerated fluxes:	
Burning of fossil fuels (**F** to **A**)	5
Deforestation (**P** to **A**)	2

*Note that the carbon budget does not quite balance; carbon fluxes are not well enough known to account for all of the carbon thought to enter the atmosphere. The unknown carbon "sink" may be land plants and soils.

Sources: Houghton, J., Jenkins, G. and Ephraums, J., 1990. *Climate Change: The IPCC Scientific Assessment.* Cambridge Univ. Press (for UN Intergovernmental Panel on Climate Change). National Research Council, 1986. *Global Change in the Geosphere-Biosphere.* Washington: National; Academy Press.

Pencil "leads" are made of graphite -- a crystalline form of carbon.

Diamonds are also crystalline carbon, but form only at the high pressures typical of the Earth's mantle.

A new form of crystalline carbon was discovered in 1990. Soccer-ball shaped carbon molecules named fullerenes, have unusual electrical and optical properties and may find industrial uses.

HCl is the main ingredient in household limescale removers.

acidic solution: aqueous solution with a higher concentration of hydrogen (H^+) than hydroxide (OH^-) ions

chemical weathering: decomposition of rocks by chemical processes

5. Carbon in rocks and minerals

a) Examine the specimens of carbon-bearing geologic materials, and briefly describe their physical characteristics and practical uses. Consider also the parts these materials play in the global carbon cycle (Fig. 5-4).

b) Place a small amount of dilute hydrochloric acid (HCl) on a sample of limestone. What happens?

This indicates that a gas is being given off as HCl reacts with the rock. What gas is it? Remember that limestone consists mainly of the mineral calcite ($CaCO_3$).

B. Local carbon fluxes

Though less visible to us than the cycling of water, carbon too is constantly being transferred between reservoirs on a local scale.

1. Carbon released by chemical weathering of carbonates

When you put HCl on the limestone specimen, you demonstrated that <u>acidic</u> solutions cause <u>chemical weathering</u> of carbonate rocks and liberate the ancient atmospheric carbon dioxide stored in them. Rainwater is generally acidic -- even in unpolluted areas -- because it incorporates atmospheric carbon dioxide as it falls, forming carbonic acid:

$$Water \ + \ Carbon \ Dioxide \ = \ Carbonic \ Acid \ in \ solution$$
$$H_2O \ + \ CO_2 \ = \ H^+ + HCO_3^-$$

The acidity of a solution is described by its <u>pH</u>, a measure of the concentration of hydrogen ions (pH stands for potential of hydrogen). Neutral solutions have pH values of 7, acids values less than 7 and bases values between 7 and 14 (Fig. 5-5).

FIGURE 5-5.
PH VALUES
FOR COMMON
SOLUTIONS

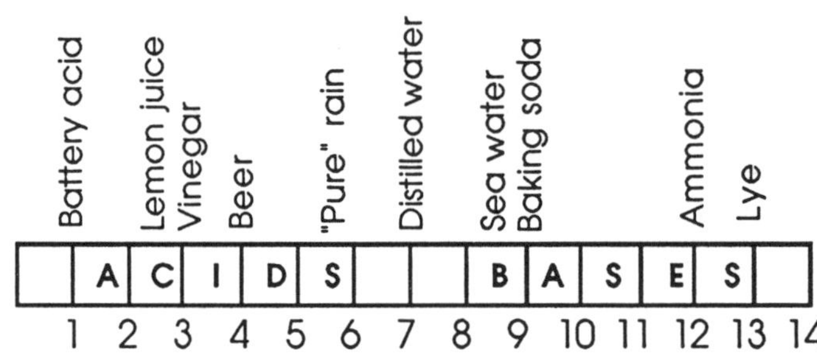

This is how most caves are formed.

By taking up atmospheric carbon dioxide in solution, rainwater may reach a pH of 5.5. It can become a still stronger acid after incorporating CO_2 from organic matter in the soil. When this acidic water percolates into the subsurface and comes into contact with limestone bedrock, some of the limestone dissolves into calcium and bicarbonate ions, which are then carried in solution by surface and groundwater to the sea (see Fig. 5-5).

$$Limestone \; + \; Hydrogen \; ions \; = \; Calcium \; + \; Bicarbonate$$
$$CaCO_3 \; + \; H^+ \; = \; Ca^{++} \; + \; HCO_3^-$$

Acid rain is formed when water in the atmosphere combines with pollutants like sulfur dioxide (SO_2, formed in the burning of sulfur-rich coal) and nitrogen oxide (NO_x, from car exhaust). See Lab 7.

Like antacid tablets, which consist largely of $CaCO_3$, limestone buffers, or neutralizes, acids.

a) Following the instructions provided, use litmus paper or a pH meter to measure the acidity of precipitation collected in your area.

pH of rainwater (or melted snow) = ___________

Is it acidic? More acidic than you would expect unpolluted precipitation to be? (See Fig. 5-5).

Put several pieces of limestone into the water and measure the pH again:

pH immediately after adding limestone = ______________

Wait 2 minutes and measure the pH a third time:

pH 2 minutes after limestone was added = ______________

How can you explain the changes in the pH of the water?

b) Determine the pH of the following local water samples, if available.

Spring water: _______   Well water: ________   River water: ________

Are the pH values of water from these sources different from that of rainwater?

c) What is the bedrock in your area composed of?
Does it appear to buffer (neutralize) the acidity of precipitation? Explain.

2. Carbon dioxide released by burning of fossil fuels
If you drive a car, heat your house with natural gas, or use electricity generated at a coal-burning power plant, you contribute a significant amount of carbon dioxide to the atmosphere each year. Fossil fuels including coal, oil and natural gas formed millions of years ago from plant matter that was buried in sediments before it could decay. Like modern vegetation, these plants extracted carbon dioxide from the atmosphere to carry out photosynthesis. When fossil fuels are burned, this carbon is returned to the atmosphere.

a) How much CO_2 is generated each year by fossil fuel use? To determine this, first turn to Table 5-4. How much carbon is released yearly by fossil fuel burning?

_______________ metric tons (1000 kg or 2200 lbs)

Next, find the <u>atomic weights</u> of carbon (C) and oxygen (O) in a periodic table of the elements.

    Atomic weight of C: _______   Atomic weight of O: _______

So what is the weight of a molecule of CO_2, consisting of one carbon and two oxygen atoms?

How many times heavier is a molecule of CO_2 than an atom of C?

How many tons of CO_2 are released annually by fossil fuel burning?

b) A gallon of gasoline produces about 9 kg (20 pounds) of CO_2 when combusted. How many gallons of gas do you use in a week? (If you don't drive a car, use a value of 10 gallons/week). How much CO_2 does that produce? Show your calculations.

How much CO_2 does your car emit in the course of a year?

atomic weight:
a dimensionless number expressing the relative weight of an atom of an element. The atomic weight of the lightest element, hydrogen, is about 1.

Wood burning, like fossil-fuel burning, contributes CO2 to the air. The only difference is that the carbon in the wood has been stored for decades rather than millions of years.

greenhouse gas: atmospheric gas that is transparent to incoming sunlight but blocks reflected heat energy from returning to space

Over time, small imbalances in carbon fluxes can have significant effects.

In 1986, sudden release of volcanic CO_2 that had accumulated in a lake in Cameroon, West Africa asphyxiated more than 1700 people.

C. Climate and the carbon cycle

Many types of geologic evidence indicate that the Earth's climate has alternated between warmer and colder periods in the past. Glacial deposits in the mid-continent US record the most recent period of global cooling about 18,000 years ago.

The causes of these climatic fluctuations are not fully understood, but there is strong evidence that periods of relatively high global temperatures have been associated with high levels of atmospheric carbon dioxide (CO_2). Carbon dioxide is a <u>greenhouse gas</u> -- that is, it allows incoming short-wavelength light energy from the sun to pass through the atmosphere but does not allow longer-wavelength heat energy that is reflected back from Earth to return to space. Heat consequently becomes trapped near the surface of the planet. A similar effect occurs in a greenhouse or in a closed car on a sunny day. The planet Venus, with its thick, CO_2-dominated atmosphere (Lab 1), is a runaway greenhouse.

1. Long-term carbon storage on Earth

Review Table 1-2 in Laboratory 1. How much CO_2 is in Venus' atmosphere (as a percent)? In Earth's atmosphere?

What process in the Earth's carbon cycle has apparently kept Earth from becoming a runaway greenhouse? Hint: What is the largest carbon reservoir on Earth? Review Table 5-3 and your answer to question II.A.3.

Although Earth has only trace amounts of CO_2 in its atmosphere compared with Venus, small variations in the concentrations of Earth's atmospheric CO_2 have had dramatic effects on climate. What has caused natural fluctuations in atmospheric carbon dioxide? A glance at the diverse components of the carbon cycle (Tables 5-3 and 5-4) suggests that many processes could be involved...

2. Carbon and tectonics

On the time scale of days and weeks, carbon cycling occurs mainly on and above the Earth's surface as carbon is exchanged between air, water, soil, and living organisms. Over the longer term, however, the carbon cycle is controlled by processes deep within the Earth.

a) Which of the fluxes listed in Table 5-4 are directly related to tectonic processes?

Do these tend to increase or decrease the amount of CO_2 in the atmosphere?

The unusually rapid rates of seafloor spreading in the Cretaceous meant that most of the world's ocean crust was young, hot and isostatically buoyant. As a result, sea level rose and vast areas of the continents were flooded.

The late Cretaceous period was also unusual in that no magnetic polarity reversals (Lab 3) occurred for a period of more than 30 my, from 118 -84 my ago.

The dinosaurs thrived in the warm Cretaceous climate, occupying even Antarctica and Northern Alaska, which lay close to their present latitudes.

FIGURE 5-6.
DEEP SEA TEMPERATURES FOR THE PAST 100 MILLION YEARS
Temperatures estimated from oxygen isotope ratios in shells of marine creatures preserved in sea sediments.

b) Seafloor spreading and climate
Based on your answer to the previous question, how do you think climate might be affected by an increase in global rates of sea floor spreading ?

In Laboratory 3 (question III.A.2, p. 53), you estimated the rates of seafloor spreading for the Pacific and Atlantic oceans over the past 65 million years (the Cenozoic Period). What was the rate that you determined for the north Atlantic ridge over this time period?

_______________ cm/yr

Using Figure 3-4 and the procedure you followed before, determine the average rate of seafloor spreading on the Atlantic ridge at the latitude of southern Florida for the Cretaceous period (65 to 144 million years ago). Be sure to measure the widths of Cretaceous ocean crust both east and west of the ridge. The map scale is 1 cm = 3200 km.

$$\frac{\text{Total map width of ocean crust (cm)}}{\text{Time period in years}} \quad \times \quad \frac{\text{Map scaling}}{\text{factor}} \quad = \quad \frac{\text{Spreading rate}}{\text{in cm/yr}} \qquad (5\text{-}4)$$

How much faster was seafloor spreading on this part of the Atlantic ridge during the Cretaceous compared with spreading during the Cenozoic?

Unusually rapid seafloor spreading rates occurred on all of the mid-ocean ridges during much of the Cretaceous Period and had a variety of global consequences. Figure 5-6 illustrates seawater temperatures for the past 100 million years. When were temperatures highest? Does it appear that your prediction about the climatic effects of rapid spreading was correct?

What is the overall trend in temperatures since the Cretaceous? When was there a significant reversal in this trend?

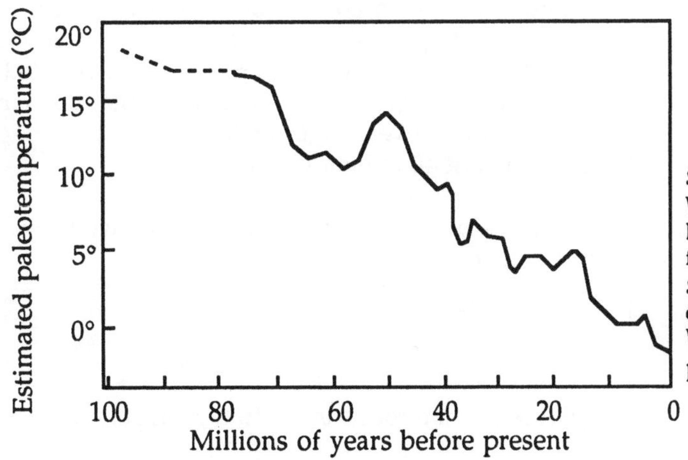

Source: Douglas, R. & Woodruff, F., 1981. Deep-sea benthic foraminifera, in *The Sea*, v. 7, C. Emiliani, editor. New York: Wiley. Used with permission.

c) Metamorphism, mountain building and climate

Precipitation of carbonate rocks like limestone extracts CO_2 from Earth's atmosphere and keeps it out of the carbon cycle for long periods of time. Eventually, though, this carbon may be returned to the atmosphere (via faults in the Earth's crust) if the carbonate rocks undergo metamorphism at a convergent plate boundary. Metamorphism of limestone (in the presence of some silica or quartz) can be written as a simple chemical equation:

$$CaCO_3 \quad + \quad SiO_2 \quad = \quad CaSiO_3 \quad + \quad CO_2$$

(Calcite, the main   (Silica   (Wollastonite,   (Carbon
mineral in limestone) or quartz)   a silicate mineral)   dioxide)

What effect would widespread metamorphism of carbonates have on climate?

In a sense, chemical weathering and metamorphism can be considered converse processes. Weathering extracts volatiles like H_2O and CO_2 from the atmosphere and binds them to minerals; metamorphism returns these components to active circulation.

Another carbon flux that is related to convergent tectonism is the rate of rock weathering. As mountains grow during continental collision, erosion rates accelerate in response to the steeper topography, increasing the amount of surface area available for chemical weathering. As you discovered when you put hydrochloric acid on a piece of limestone, chemical weathering of carbonate rocks releases CO_2 into the atmosphere. However, chemical weathering of *silicate* rocks, which make up most of the continental crust, *extracts* CO_2 from the air through reactions like:

$$CaSiO_3 \quad + \quad CO_2 \quad = \quad CaCO_3 \quad + \quad SiO_2$$

A representative   (dissolved in   Calcite   Silica
silicate mineral   rainwater)   (dissolved in   (quartz)
river & sea water)

How is this chemical equation related to the previous one?

In the building of great mountain belts like the Himalaya, the main period of metamorphism of carbonate rocks typically occurs early, as subduction carries marine rocks to great depth. The creation of steep mountain topography occurs later, when the two continental masses have collided and subduction has ceased. In the case of the Himalaya, the main pulse of metamorphism occurred between 45 and 55 million years ago while the high and rugged topography was created only in the last 40 million years.

Given this sequence of events, what effects would you expect the formation of the Himalayan chain to have had on climate?

Is your prediction consistent with the temperature trends shown in Figure 5-6?

3. Carbon dioxide and climate change in the geologically recent past
Clearly, tectonic processes may be important in dictating the long-term changes in Earth's climate. But growing concern about human effects on Earth's atmosphere and climate make it important to understand the causes of shorter-term changes in the past. In a sense, we need to extend the idea of Uniformitarianism -- the principle that the present is the key to the past (Lab 2) -- and use records of the past as the key to the *future*.

So how can we reconstruct atmospheric compositions at times in the past? We need to find places where "old air" has been preserved. Ice caps are such archives of old air. As yearly snows accumulate and compact, air in underlying layers is trapped. By collecting and analyzing cores from long-lived ice caps, detailed information about past atmospheric conditions can be retrieved. Cores taken at the Vostok research station in Antarctica and at sites near the summit of the Greenland ice cap provide records of atmospheric chemistry back to 200,000 years before present -- well before the most recent ice age. The layers in these ice cores are rather like the annual rings in a tree, and they reveal that Earth's climate has undergone fluctuations not only on the scales of hundreds of thousands and tens of thousands of years but even over centuries and decades. Figure 5-7 shows carbon dioxide and temperature data from the Vostok ice core.

a) Describe the relationship between the plots of temperature and CO_2 data from the Vostok ice core (Fig. 5-7). Do the maximum and minimum values in the two occur at the same times?

Ice cores provide the most detailed records of changes in Earth's atmosphere and climate. Other natural sources of climatic information include tree rings, lake sediments, deep-sea sediments and corals, all of which grow or accumulate at known rates.

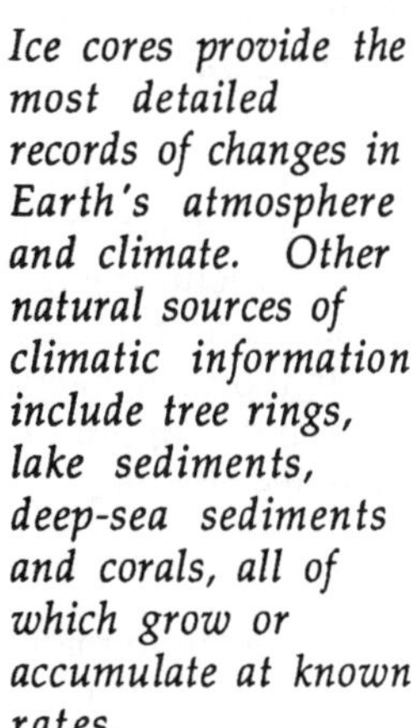

FIGURE 5-7.
CARBON DIOXIDE AND TEMPERATURE DATA FROM THE VOSTOK ICE CORE
Temperatures estimated from measured concentrations of deuterium (a stable isotope of hydrogen).

FIGURE 5-8.
EXPECTED CHANGES IN EARTH'S SURFACE TEMPERATURE ASSOCIATED WITH MILANKOVITCH ORBITAL CYCLES

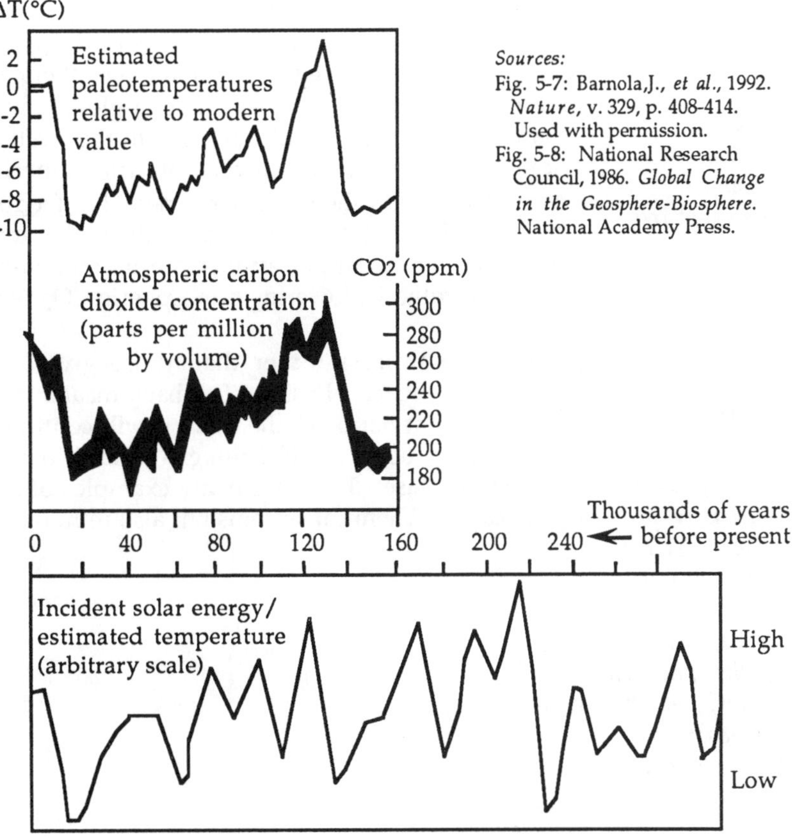

Sources:
Fig. 5-7: Barnola,J., *et al.*, 1992. *Nature*, v. 329, p. 408-414. Used with permission.
Fig. 5-8: National Research Council, 1986. *Global Change in the Geosphere-Biosphere.* National Academy Press.

feedback mechanism: a process in which the output (effect) of one phase becomes the input (cause) of the next

Milankovitch orbital cycles: cyclical variations in Earth's orbit and rotation that may affect the amount of solar energy the planet receives

positive feedback mechanism: a self-perpetuating process

negative feedback mechanism: a self-correcting process

A familiar example of positive feedback is the shrieking produced when a stereo speaker broadcasts a tone detected by a microphone, which picks up the tone from the speaker, which then broadcasts a still louder tone...

b) Orbital cycles, carbon, and climate

The causes of the fluctuations in atmospheric CO_2 over the past 200,000 years are not well understood. In some cases, the CO_2 levels actually seem to lag slightly *behind* the temperature record (Fig. 5-7), suggesting complex <u>feedback mechanisms</u> between temperature and CO_2 levels. That is, some other factor may have caused changes in Earth's surface temperature, which then lead to changes in atmospheric CO_2 (which may have further affected temperature...).

Many people who study climate change believe that recent cycles of global warming and cooling, particularly the Pleistocene ice ages, may have been triggered by cyclical changes in Earth's orbital characteristics, including the shape of its path around the Sun, the tilt of its rotational axis, and the slow wobble of that axis. These orbital changes are often called <u>Milankovitch cycles</u> for the Serbian scientist who first suggested that they could change the amount of solar energy reaching the Earth and therefore influence the planet's climate.

Compare Figures 5-7 and 5-8. Do the data from the Vostok ice core seem to match the Milankovitch predictions? Which parts of the temperature and CO_2 plots coincide most closely with the orbital cycles? Which parts differ?

Assuming Milankovitch orbital cycles did cause some of the observed fluctuations in Earth's surface temperatures, how might the carbon cycle have responded? Some of the responses may have involved <u>positive feedback</u> mechanisms -- that is, temperature-induced changes in atmospheric CO_2 that further accentuated the temperature change. For example, a global temperature increase may result in thawing of swampy tundra regions in the Arctic, which accelerates the decay of organic matter, raises levels of atmospheric CO_2 and further increases temperature. On the other hand, <u>negative feedback</u> mechanisms -- carbon-cycle changes that act to 'damp' or reduce the temperature changes -- are also plausible. For example, an increase in temperature could accelerate global rates of photosynthesis, lowering atmospheric CO_2 levels and cooling Earth again.

Describe a process (other than those given) that exhibits either positive or negative feedback. Positive feedback means that the process is *self-perpetuating* -- the more it happens, the more it will continue to happen. Negative feedback means the process is *self-limiting* -- once it happens, it results in conditions that cause it to cease. There are many examples of such processes not only in natural physical and chemical systems but also in human behavior and interactions.

D. Human effects on climate
If Earth's climate has varied significantly in the past, why is there concern about how human activities may alter it? The reason is that humans seem to be changing atmospheric chemistry at rates that far exceed the rates of natural changes -- so records from the past may not allow us to predict the consequences.

1. Increases in atmospheric CO_2 increase since the Ice Age
a) Look again at Figure 5-7. What was the concentration of atmospheric CO_2 (in parts per million by volume) at the height of the last ice age 18,000 years ago?

b) Figure 5-9 shows levels of atmospheric carbon dioxide since the time of the industrial revolution. What was the CO_2 concentration in 1800?

How much did the concentration of atmospheric CO_2 increase between the height of the last ice age and the beginning of the industrial revolution?

Although it is the most abundant volumetrically, CO_2 is not the only green-house gas. Methane (CH_4), exhaled by swamps, landfills, and ruminant animals like cows, can trap more heat on a molecule-by-molecule basis than CO2.

What was the average *rate* of increase during this time (in ppm/year)?

What is the present concentration of atmospheric CO_2? How much has it increased since the beginning of the industrial revolution?

What was the average rate of increase during this time?

c) What was the highest concentration of atmospheric CO_2 during the interglacial period about 125,000 years ago (Fig. 5-7)?

How much warmer were temperatures then than they are today?

FIGURE 5-9.
ATMOSPHERIC CARBON DIOXIDE SINCE 1700
Based on ice core and atmospheric samples.

Source: Friedli, H., *et al.*, 1986. *Nature* , v. 324, p. 237-238. Used with permission.

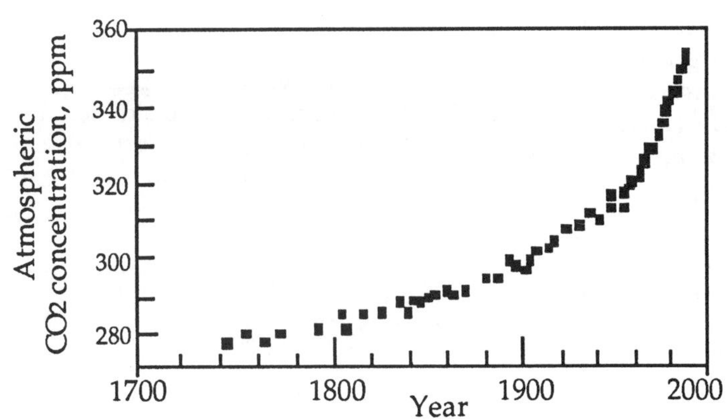

Even a small rise in global sea level would threaten significant portions of the world's food producing areas, especially low-lying rice paddies that sustain millions in the Eastern Hemisphere.

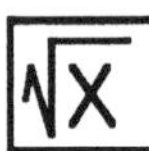

Rising sea level could also degrade ground-water as a result of increased salt content and the spread of contaminants from flooded garbage dumps.

In Holland, a country that lies very near sea level, an elab-orate system of dikes and pumps keeps the sea from flooding the land. The 1990 cost of a 1 km dike segment was US $1 billion.

Increased hurricane frequency is another potential consequence of higher global temperatures. Hurri-canes are spawned where sea surface temperatures are unusually high.

The total cost of hurricane Andrew in 1992 exceeded $15 billion.

2. Climate change and the oceans

There is a lag time between increases in atmospheric CO_2 and increases in global temperature because it takes time for the oceans to warm up. For now, the oceans are keeping us cool. There is disagreement about the magnitude of the temperature increase that will result from human production of greenhouse gases, but most climate modelers agree that temperatures will begin to rise perceptibly by early next century. Even seemingly insignificant increases in average global temperature can have surprisingly large and varied effects (see Table 5-5). One of the likely effects is rising sea level.

Earlier in this laboratory, you estimated the sea level rise associated with glaciers melting at the end of the last ice age. Another factor contributing to sea level rise is the thermal expansion of water -- that is, warm water takes up more space than cold.

a) Use the equation below to determine how much sea level might rise due to thermal expansion alone if global temperatures were to reach the value estimated for the last interglacial period (question **c** on previous page):

| Coefficient of thermal expansion | x | Temperature increase | x | Ocean volume | / | Ocean area | = | Sea level rise | (5-5) |

$$2.1 \times 10^{-4}/°C \quad \times \quad \underline{\quad}°C \ \times \ 1.2 \times 10^9 \, km^3 \ / \ 3.5 \times 10^8 \, km^2 \ = \ \underline{\quad} km$$

$$= \ \underline{\quad} m$$

Current levels of atmospheric CO_2 are already higher than they were 125,000 years ago. Calculate the potential sea level rise if the oceans were to warm by 5°:

b) In either scenario, what major world cities would face coastal flooding and erosion even if there were no further melting of glaciers and ice caps?

What do you advocate as the best policy governments can implement to address the problems of warming-induced sea level rise and increased hurricane frequency? What strategies do you think would make the best use of your tax dollars? If you live in a coastal area, cite local issues to justify your stance.

<table>
<tr><td>

TABLE 5-5.

POSSIBLE EFFECTS OF GLOBAL WARMING ON ECONOMIC RESOURCES

Source: Smith, J. and Mueller-Vollmer, 1992. "Setting priorities for adapting to climate change". Contractor paper for Office of Technology Assessment, US Congress.

Analysis of cores collected in 1993 from the summit of the Greenland ice cap indicates that the relative stability of Earth's climate in the last several centuries may be the exception rather than the rule. Climate modelers predict that Earth's future climate may be ... unpredictable.

</td><td>

<u>RESOURCE</u>	<u>POTENTIAL EFFECTS</u>
Water	Changes in supply due to shifting precipitation patterns Changes in drought and flood frequency Changes in water quality
Coastal property	Increased losses to erosion Increased risk of flooding Increased risk of severe storms
Forests	Migration of vegetation types Reduction in range Altered ecosystem composition
Agriculture	Changes in crop yields Geographic shifts in productivity
Energy	Increase in cooling demand Decrease in heating demand Changes in hydropower output
Transportation	Risks to coastal roads Fewer disruptions of winter transport
Human health	Shifts in ranges of infectious diseases Changes in heat-stress and cold-weather related afflictions

</td></tr>
</table>

III. Synthesis

A. Preview review

Return to the Lab Preview and complete any questions you were unable to answer before.

B. Personal water budget

Estimate how much water you use in a typical day, using the values given below for water consumed in everyday activities. List other ways in which you directly or indirectly use water every day (e.g. you eat vegetables and grains that were irrigated; use paper that required vast amounts of water to manufacture...). Because of acute water shortages in many parts of the world, 2 billion people use less water each day than we use each time we flush the toilet.

Tub bath	30-40 gallons	Flushing standard toilet	3 gallons
Average shower	20-30 gallons	Machine washing 1 load of clothes	20-30 gallons

C. Icehouse memories

How was your region affected by the most recent Ice Age? Write a short report describing local geologic features that record this period of global cooling.

D. Greenhouse news

Find a recent article about some aspect of climate change in a newspaper or magazine and write a short analysis of it based on what you have learned in this laboratory. You might choose a topic from Table 5-5.

SUGGESTED REFERENCES AND RESOURCES

Water:

National Geographic, November 1993. *Water* (Special Issue).
>   A far-ranging exploration of water issues, with a special section on the 1993 Mississippi floods. Includes a map supplement showing rainfall patterns and major aquifers in the US.

Gleick, P. H., ed. *Water in Crisis: A Guide to the World's Fresh Water Resources.* Oxford: Oxford Science Publications, 573 pp.
>   A valuable compendium of facts and figures, addressing water availability and quality throughout the world.

Moustafa, T. C., 1992. The hydrological cycle and its influence on climate. Nature, v. 359, p. 373-379.
>   A useful review article arguing that inadequate understanding of the hydrologic cycle is a major limitation in climate modeling.

Carbon cycle/climate change:

Bell, M., 1994. Is our climate unstable? Earth, v. 3 n. 1, p. 24-31.
>   A good summary of preliminary interpretations of data from ice cores collected in 1993 at the Greenland ice cap summit. The cores record dramatic decade-scale fluctuations in climate in the last interglacial period and suggest that the present interval of climatic stability is an anomaly.

Houghton, J. T., Jenkins, G.J. , and Ephraums, J.J., eds. 1990. *Climate Change: The Intergovernmental Panel on Climate Change Scientific Assessment.* Cambridge: Cambridge University Press, 365 pp.
>   An authoritative document prepared for the UN Environment Program and World Meteorological Organization. An updated edition is due late in 1994.

Nordhaus,W., 1994. Expert opinion on climate change. *Am. Scientist*, v. 82, p. 45-51.
>   Summary of a survey of leading scientists and economists on climate change. Reveals the wide range of views held even by "experts".

Office of Technology Assessment, 1993. Preparing for an Uncertain Climate (2 vols). Washington DC: US Government Printing Office. Pub. no. OTA-0-567 & -568
>   An overview of scientific and policy issues related to human-accelerated global change.

Schaufele, C. and Zumoff, N., 1994. Earth Algebra. Harper Collins Publishers.
>   A ground-breaking mathematics text that introduces algebraic concepts through questions and exercises related to carbon dioxide emissions and climate change.

High Performance Systems, Inc., 1993. *STELLA II* v. 3.0 Macintosh software.
>   Software to facilitate quantitative analysis of complex systems and cycles that would otherwise require use of differential equations. First designed for studies of population dynamics in ecosystems, STELLA is ideal for explorations of geochemical cycles. Users graphically define systems in terms of reservoirs ("stocks") and fluxes ("flows"), then can explore how changes in these affect system equilibrium over time. HPS Systems, Inc. 45 Lyme Road, Suite 300, Hanover, NH 03755. Phone: 603/643-9636; FAX 603/643-9502.

THE ORGANIC EARTH

How have the characteristics of Earth influenced life?
How has life influenced the characteristics of Earth?

In the preceding laboratories, you have explored some ways in which Earth is different from nearby planets. Earth has an atmosphere of unusual composition, an active hydrologic cycle, and a lithosphere that is in continuous motion. Most significantly, of course, Earth is home to the only known life forms in the Solar System.

Is it a coincidence that Earth has all of these peculiar attributes? Or are they somehow linked to one another? Did life come to Earth because the conditions happened to be right? Or did life play an active part in shaping the conditions in which it would live?

OBJECTIVES

- To explore how the biosphere is linked with the atmosphere, hydrosphere and solid Earth
- To examine chemical cycles on Earth and consider how they would be different in the absence of life

OUTLINE

I. Raw materials for life
> Chemistry of the human body
> The phosphorous cycle
> The nitrogen cycle
> Links among the phosphorous, nitrogen and other chemical cycles

II. History of Earth's atmosphere
> Clues from other planets
> Clues from ancient rocks

III. The Gaia hypothesis
> Daisyworld
> Earthly examples

<u>LAB PREVIEW</u> Before coming to lab,
- Answer as many of the following questions as you can
- Read the following chapters in *The Blue Planet:*
    Chapter 15: "Dynamics of Global Ecosystems"
    Chapter 16: "Evolution of the Biosphere"
Or other readings assigned by your instructor

How much do you already know about life on Earth?

What are the main elements that make up the human body? From what sources do humans obtain these elements?

What elements critical for life are among the most difficult to obtain?

What was Earth's early atmosphere like? How and when did it begin to take on its present composition? What is the evidence for this?

What is meant by the Gaia hypothesis? What are your views about the hypothesis?

EXERCISES

I. Raw materials for life

Although we tend to think of ourselves as distinct from our environment, our bodies are literally made of components derived from the air, water, and rocks around us. The food we eat and water we drink link us with the atmosphere, hydrosphere and solid Earth.

A. Chemistry of the human body

How does the composition of our bodies compare with those of the atmosphere, oceans and crust of the Earth? The table below summarizes the chemistry of each. The similarities and differences in the compositions of these chemical reservoirs suggest interesting questions about the role of life on Earth.

TABLE 6-1.
COMPARATIVE COMPOSITIONS OF EARTH'S CRUST, ATMOSPHERE, SEAWATER AND THE HUMAN BODY
Elements listed in order of increasing atomic number. Values given as percent by weight.

Element (symbol, atomic number)	Human body[1]	Earth's crust[2]	Atmosphere[3]	Seawater[3]
Hydrogen (H, 1)	10.00%	0.14%	Trace	11.00%
Carbon (C, 6)	18.00	0.020	0.01	Trace[4]
Nitrogen (N, 7)	3.00	Trace	78.10	Trace[4]
Oxygen (O, 8)	65.00	46.60	20.90	88.30
Sodium (Na, 11)	0.15	2.83	Trace	1.08
Magnesium (Mg, 12)	0.05	2.09	Trace	0.13
Aluminum (Al, 13)	Trace	8.13	Trace	Trace
Silicon (Si, 14)	Trace	27.72	Trace	Trace
Phosphorous (P, 15)	1.00	0.11	Trace	Trace[4]
Sulfur (S, 16)	0.25	0.026	Trace	0.09
Chlorine (Cl, 17)	0.15	0.013	Trace	1.94
Argon (Ar, 18)	Trace	Trace	0.93	Trace
Potassium (K, 19)	0.35	2.59	Trace	0.04
Calcium (Ca, 20)	1.50	3.63	Trace	0.04
Titanium (Ti, 22)	Trace	0.44	Trace	Trace
Iron (Fe, 26)	0.004	5.00	Trace	Trace
Zinc (Zn, 30)	0.0004	Trace	Trace	Trace

[1]Cotterill, R., 1985. *Cambridge Guide to the Material World*. Cambridge Univ. Press.
[2]Mason, B., 1966. *Principles of Geochemistry*. Wiley.
[3]Brownlow, A., 1979. *Geochemistry*. Prentice-Hall. Due to uncertainties, values total >100%.
[4]Values are for the open ocean; Biological processes locally increase P,N and C concentrations.

1. Make a plot to illustrate the data given in Table 6-1. List the elements in order of increasing atomic number on the horizontal axis, and make the vertical axis a linear scale from zero to one hundred percent. Your plot could be a scatter diagram with four different symbols for the human body, crust, atmosphere and seawater, or it could be series of bar graphs for each element, each with four bars for the four columns in the table. If computers are available, make a 3-dimensional bar graph. Spreadsheet instructions are provided in Appendix I (p. 173).

Although the total concentration of elements like sodium (Na) or zinc (Zn) in the human body is low, these metals are essential for health....

2. What single element is relatively abundant (>20%) in each of the 4 'reservoirs'? In each case, what is the main form (compound) in which this element occurs? This should be obvious for seawater; use similar logic for the human body. For the atmosphere, review Table 1-2, Lab 1; for the crust, see Table 3-1, Lab 3.

...At higher concentrations, however, these elements can be health hazards; sodium can cause high blood pressure; zinc becomes toxic.

3. Other than hydrogen (which we take in by drinking water), which elements are at least 5 times more abundant in the human body than in Earth's crust?

For the elements you listed, suggest sources from which humans could obtain them.

limiting nutrient: element or molecule essential for life but available in limited quantities; if more were available, an organism's growth would be enhanced

B. The phosphorous cycle

Although humans and all other living things are made of elements derived from the atmosphere, hydrosphere and solid Earth, the differences in the composition of the human body and these other reservoirs suggests that there must be processes that collect and concentrate the raw materials necessary for life. Two of the most important life materials are the elements phosphorous and nitrogen (P and N). Both are <u>limiting nutrients</u> for many organisms; that is, of all elements necessary for life, these are in shortest supply.

Bone is a composite of the mineral apatite and the organic substance collagen. Apatite gives bone its rigidity, while collagen gives it flexibility. Teeth are also made largely of apatite.

In both plants and animals, phosphorous is an essential component of cell walls and of the genetic material DNA, yet it occurs in only trace amounts in Earth's crust. How does the biosphere obtain the phosphorous it needs? Phosphorous-bearing minerals in igneous rocks are the ultimate source of the phosphorous used by the biosphere. These minerals, particularly apatite [$Ca_5(PO_4)_3OH$], release their phosphorous as they undergo weathering at the Earth's surface. This process is so slow, however, that it is important only over the time scale of hundreds of millions of years. Most of the phosphorous taken up by the biosphere is "recycled" -- i.e., phosphorous that has already been used many times by organisms. Like the water and carbon cycles (Laboratory 5), the phosphorous cycle can be characterized in terms of reservoirs, fluxes, and residence times. The main components of the global phosphorous cycle are summarized in Tables 6-2 and 6-3.

TABLE 6-2.
PHOSPHOROUS
RESERVOIRS
ON EARTH

Reservoir	Mass in billions of kilograms
Oceans (O)	
Dissolved in deep ocean water	100,000
Biosphere/Land (B_L)	2000
Land plants and animals	
Biosphere/Ocean (B_O)	140
Ocean biota	
Soils (S)	
(Not all P in soils is in forms accessible to plants)	150,000
Sedimentary rocks (R)	
Submarine phosphate deposits	9×10^8
Mineable phosphate deposits (exposed above sea level)	19,000

TABLE 6-3.
PHOSPHOROUS
FLUXES
ON EARTH

Natural fluxes	Billions of kilograms/year
Uptake of phosphorous from soil by land plants (S to B_L)	250
Decomposition of land plants and animals (B_L to S)	250
P from soil carried by rivers to ocean, taken up by biota (S to B_O)	1.5
Decay of ocean biota (B_O to O)	120
Upwelling of deep-ocean organic P to shallow depth (O to B_O)	120
Deposition of phosphate sediments on ocean floor (O to R)	0.5
Weathering of tectonically uplifted phosphate sediments (R to S)	0.5
Human-accelerated fluxes	
P mined from phosphate sediments for fertilizer, detergent (R to S)	14
Excess P in river water from soil erosion, fertilizer runoff, and detergents (S to O)	5.5
P removed from oceans by marine fishing (B_O to B_L)	50

Source: National Research Council, 1986, *Global Change in the Geosphere-Biosphere.* Natl. Academy Press.

THE
PHOSPHOROUS
CYCLE ON EARTH

1. Using the information in Tables 6-2 and 6-3, draw a schematic diagram to illustrate the phosphorous cycle. Begin by drawing the reservoirs, then add labeled arrows whose widths reflect the relative flux rates into and out of these reservoirs. Use Figures 5-1 and 5-4 (Lab 5) to help you get started.

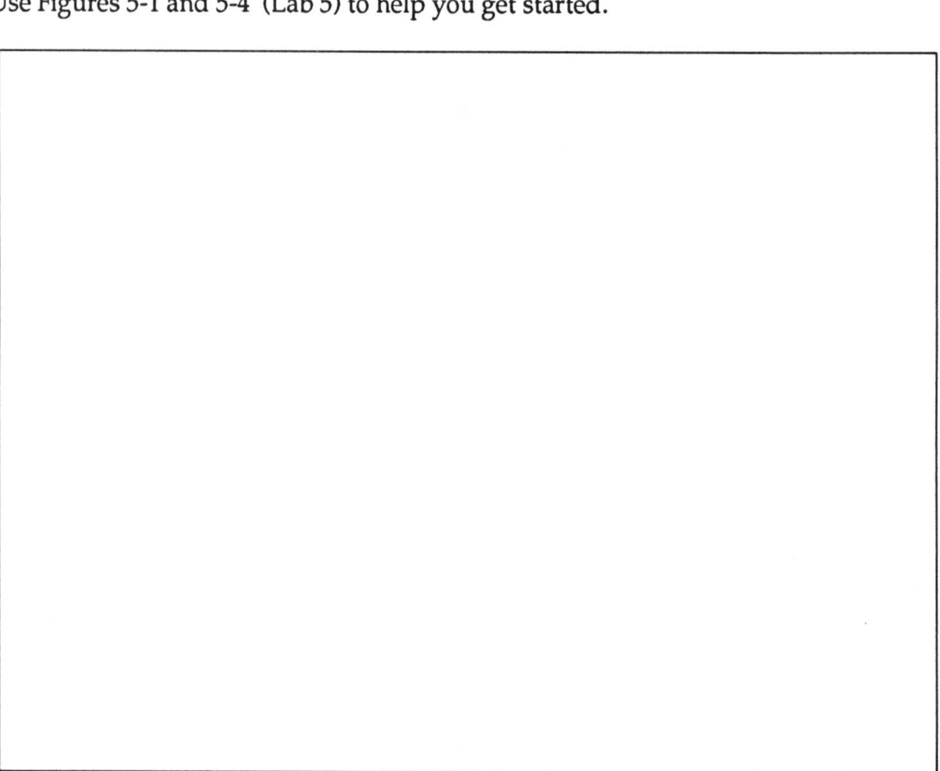

2. Examine the specimens of phosphorous-bearing minerals, rocks, and biological materials and briefly describe their characteristics (e.g. color, texture, hardness).

phosphorite: chemical sedimentary rock made of the mineral carbonate apatite $Ca_5(PO_4,OH,CO_3)_3(F)$ typically deposited in low latitude, shallow marine environments where upwelling currents bring deep-ocean phosphorous to surface waters

The value of phosphorous as a fertilizer has been recognized for centuries. Native Americans buried fish skeletons with seeds to enhance plant growth; English farmers applied ground bone to their crops.

Guano (bird droppings) is also rich in phosphorous and was an internationally traded commodity in the 19th century.

3. What is the largest phosphorous reservoir on Earth (other than the crust, the ultimate source of all phosphorous at Earth's surface)?

What is the principal source of phosphorous for plants and animals on land?

What is the principal source of phosphorous for plants and animals in the sea?

In what way are these land and sea reservoirs similar? (How are the reservoirs replenished?)

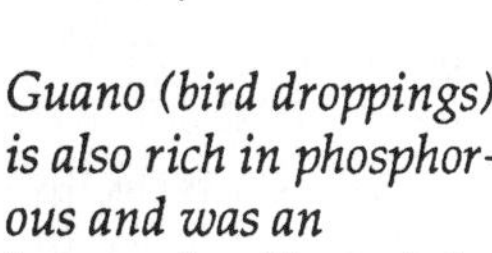

In parts of Africa where soils are very low in phosphorous, cattle, giraffes and other large mammals sometimes eat bones, apparently to obtain the phosphorous they require.

4. Why do you think soils used for agriculture tend to become depleted in phosphorous? Where does the phosphorous in cultivated crops go? How is this different from the fate of phosphorous in plants growing in the wild?

5. Why is tectonic activity important in the global phosphorous cycle?

6. How does the rate at which humans mine phosphate sediments compare with the natural rate of weathering of exposed phosphate deposits? That is, how many times faster is phosphate released by mining than weathering?

photosynthesis:
process by which green plants, with sunlight as their energy source, convert water and CO_2 to sugars and O_2

primary producers:
organisms able to derive energy from inorganic sources. Also called **autotrophs** ("self-feeders")

food chain:
a hierarchy of organisms in which energy and nutrients are transferred from primary producers to the organisms that consume them -- to the organisms that consume those organisms, etc.

THE NITROGEN CYCLE ON EARTH

nitrogen fixation:
biochemical process in which bacteria convert atmospheric nitrogen (N_2) to compounds including ammonium (NH_4^+)

denitrification:
return of fixed nitrogen -- NH_4^+, NH_3^+ (ammonia), and NO_3^- (nitrate) -- to the atmosphere as N_2

C. The nitrogen cycle

Nitrogen is another critical component in the life processes of both plants and animals. It is essential, for example, in facilitating <u>photosynthesis</u>, by which green plants manufacture sugars from atmospheric carbon dioxide and water. Most plants actually grow under conditions of nitrogen deficit; this is why nitrogen fertilizers are so effective in enhancing plant growth. In animals, including humans, nitrogen is a vital component of amino acids, the building blocks of proteins in skin, bones and muscles. Animals take in the nitrogen they need by eating plants (or other plant-eating animals). Plants are thus the <u>primary producers</u> of nitrogen for organisms higher in the <u>food chain</u>. Why is it so difficult for plants to obtain nitrogen, the most abundant constituent of Earth's atmosphere?

Though abundant, atmospheric nitrogen is locked in the form N_2, a molecule with a bond so strong that lightning is the only natural physical process that can break it. But the amount of nitrogen made available by lightning is insignificant compared to the amount needed by life on Earth. On land, the main process that converts atmospheric nitrogen to usable form is chemical <u>nitrogen fixation</u> by bacteria that live in the root systems of a limited number of plant species, most notably the legumes (peas, beans, clover, alfalfa). Some types of marine algae are also capable of nitrogen fixation. Biological nitrogen fixation is balanced by <u>denitrification</u>, the return of fixed nitrogen to the atmosphere as N_2. Nitrogen fixation and denitrification are just two components of the complex global nitrogen cycle. The main components of the nitrogen cycle are summarized in Tables 6-4 and 6-5.

Use these tables to answer the following questions.

1. Draw a schematic diagram to illustrate the nitrogen cycle, as you did for phosphorous.

TABLE 6-4.
NITROGEN RESERVOIRS ON EARTH

Reservoirs	*Mass in billions of metric tons*
Atmosphere (A)	
As N_2	4×10^6
As fixed nitrogen: NH_3^+ (ammonia) , NO_x (various oxides)	0.003
Oceans (O)	
As NO_3^- (nitrate), mainly in deep ocean	800
Soils (S)	
As organically bound nitrogen, NH_4^+, and NO_3^-	70
Biosphere/Land (B_L)	
As organic nitrogen in land plants and animals	10
Biosphere/Oceans (B_O)	
As organically bound nitrogen in ocean biota	1

TABLE 6-5.
NITROGEN FLUXES ON EARTH

Imbalances in input and output fluxes reflect uncertainties in Earth's nitrogen budget.

	Billions of metric tons/year
Natural fluxes:	
Nitrogen fixation by bacteria in roots of land plants (A to B_L)	0.18
Nitrogen fixation by blue-green algae in oceans (A to B_O)	0.015
Decomposition of land plants (B_L to S)	2.3
Decay of ocean biota (B_O to O)	.05
Uptake of organic nitrogen from soil by plants (S to B_L)	2.5
Denitrification of fixed nitrogen in soils (S to A)	0.13
Denitrification of fixed nitrogen in oceans (O to A)	0.1
Soil nitrate carried by rivers to ocean; taken up by biota (S to B_O)	0.025
Fixed nitrogen delivered to land by precipitation (A to S)	0.24
Fixed nitrogen delivered to oceans by precipitation (A to O)	0.1
Fixed nitrogen diffusing from soils to atmosphere (S to A)	0.2
Human-accelerated fluxes:	
Fixed N (NO_x) added to atmosphere by fossil fuel burning (to A)	0.2
Excess nitrate in river water from erosion, fertilizer runoff (S to O)	0.05

Source: National Research Council, 1986, *Global Change in the Geosphere-Biosphere.* Natl. Academy Press.

In the US, the wide-spread loss of soils to wind erosion during the "dust bowl" of the 1930's made soil conservation a national priority. Simple changes in farming practices have slowed physical soil loss from agricultural areas, but chemical depletion of soil nutrients is still treated mainly with artificial rather than natural fertilizers like manure or crop chaff.

Remember that residence time = reservoir size/flux rate.

The nitrate and phosphate runoff from fertilizers stimulates growth of algae and water plants in rivers and lakes. This process of eutrophication often results in fish kills, because decomposition of the algae reduces oxygen levels in the water. At very high concentrations, nitrates are simply toxic to organisms, including humans.

2. What is by far the largest nitrogen reservoir on Earth?

How many times larger is this reservoir than all the others combined? Show your calculations.

What are the second and third largest nitrogen reservoirs on Earth?

Which reservoir is the major source of nitrogen for most land plants?

By what paths does nitrogen enter this reservoir? By what paths does it leave ?

3. What is the average residence time for nitrogen in the soil?   Use the *input* fluxes for soils in your calculation.

4. Like the phosphorous cycle, the nitrogen cycle illustrates how the biosphere modifies its environment for its own benefit, in this case by concentrating an otherwise scarce element in the soil. Based on this and your knowledge of the effects of clearing vegetated land, describe two distinct reasons that deforestation and land clearing can lead to rapid loss of nitrogen stored in soils.

Why may it be impossible, on a human time scale, to restore soil nitrogen levels in deforested areas ? Hint: Review your answer to question **3** above.

D. Links among the nitrogen, phosphorous and other chemical cycles
You have discovered that the nitrogen and phosphorous cycles involve processes that link the atmosphere, hydrosphere, biosphere and solid Earth. But these chemical cycles are themselves linked in complex ways. Phosphorous, for example, is a critical component in activating the nitrogen-fixing capacity of the bacteria that provide usable nitrogen to the rest of the biosphere. Imbalances in the relative amounts of nitrogen and phosphorous in soils and waters can decrease biological productivity on land and in the sea. The phosphorous and nitrogen cycles are also intimately linked with the carbon cycle (Laboratory 5) since both P and N are critical elements in photosynthesis and plant growth.

Based on what you know about the phosphorous, nitrogen, carbon, and water cycles, write a brief description of how a change in a flux rate or reservoir size in one of these cycles could affect the budget of another of the cycles.

II. History of Earth's atmosphere
In Laboratory 1 (*The Earth in Space*) you found that the composition of Earth's atmosphere is entirely unlike those of its neighboring planets. How did Earth's atmosphere acquire such a composition? Has it always had an unusual mixture of elements? Or has its makeup changed with time?

A. Clues from other planets
1. Review Table 1-2, (Lab 1) and summarize how the composition of Earth's atmosphere is different from those of Venus and Mars. On the basis of this, what do you think Earth's early atmosphere might have consisted of?

2. Review your work on the carbon cycle in Laboratory 5 and explain briefly how carbon reservoirs and fluxes differ on Earth and Venus. In particular, which are absent on Venus? What appears to be the single greatest factor responsible for the differences in the atmospheres of the two planets?

supracrustals:
rocks formed at the surface of the Earth -- i.e. sedimentary and volcanic rocks in contrast to intrusive and metamorphic rocks)

Carbon has two stable (non-radioactive) isotopic forms: ^{12}C and ^{13}C. Photosynthesizing organisms preferentially draw ^{12}C from the atmosphere . When preserved in sediments, they leave carbon that is enriched in this lighter isotope.

prokaryotes:
single-celled organisms, including algae and bacteria, that lack nuclei

stromatolite:
finely layered rock formed by mats of single-celled organisms, often prokaryotic blue-green algae, that live on tidal flats. Stromatolites still form on beaches today.

B. Clues from ancient rocks

In Laboratory 5 you saw how cores from long-lived ice caps can be used to reconstruct small changes in the chemistry of Earth's atmosphere. Ice core records, however, extend back only about 200,000 years -- a tiny fraction of Earth's 4.5 billion year age. For information about Earth's early atmosphere, we must turn to ancient sedimentary rocks, which, like glacial ice preserve information about conditions at the surface of the Earth at the time they accumulated.

1. The oldest fossils

The earliest Earth was a not a hospitable place for life. During its first 400 million years, the planet was probably hit repeatedly by asteroids and other interplanetary debris (recall discussion of impact cratering in Laboratory 1). The Earth's crust was very mobile, and continents had not yet formed. The oldest preserved sedimentary rock sequence, the Isua <u>supracrustals</u> of Greenland, did not accumulate until 3.8 billion years ago -- more than 700 million years after the formation of the Earth.

The Isua rocks lack obvious fossils, but do contain organic carbon whose isotopic composition suggests the presence of early lifeforms that were already carrying on some type of photosynthesis. The oldest clearly preserved fossils come from 3500 million year old sequences in Western Australia and Swaziland in southern Africa. The fossils are mineralized remains of microscopic lifeforms called <u>prokaryotes</u>. These and many other ancient sedimentary sequences also include <u>stromatolites</u> -- finely layered rocks representing fossilized mats of single-celled organisms that lived on tidal flats.

a) Study the stromatolite specimen and make a sketch and brief description of it.

b) Why do you think early life forms often established themselves in the zone between high and low tide? What would be the advantages and disadvantages to a colony of microscopic organisms living in this environment?

banded iron formation:
chemical sedimentary rock composed of cm-scale layers of chert (microcrystalline silica) and iron oxide minerals including hematite (Fe_2O_3) and magnetite (Fe_3O_4)

reduction (of an element/compound):
ionization by addition of electrons

oxidation (of an element/compound):
ionization by loss of electrons, usually to oxygen atoms

FIGURE 6-1.
ATMOSPHERIC OXYGEN, BANDED IRON FORMATIONS AND RED BEDS
Source: Modified from Cloud, P. 1988. *Oasis in Space: Earth History from the Beginning.* W.W. Norton.

red beds:
clastic sedimentary rocks deposited on land and colored red-orange by oxidized iron

Blood is red only when the iron in it is exposed to oxygen

2. Changes in the air

Although photosynthesis, which produces free oxygen (O_2), may have begun more than 3.5 billion years ago, there was very little oxygen in Earth's atmosphere until about 2000 million years ago. Evidence for changing oxygen levels during the middle Precambrian comes from <u>banded iron formations</u>, chemical sedimentary rocks once precipitated from seawater in much the same way that limestone is precipitated from modern oceans. Banded iron formations no longer form today, however, and were deposited during a geologically brief period of time in the middle Precambrian, mainly between about 2.6 and 1.9 billion years ago (Fig. 6-1).

The widespread occurrence of Precambrian iron formations indicates that iron was once a significant dissolved constituent of seawater, just as sodium and calcium are now (Table 6-1). Iron is soluble in water, however, only in its <u>reduced</u> form, Fe^{2+}. The <u>oxidized</u> form, Fe^{3+}, is insoluble and quickly precipitated. Today, in the oxygen-rich environment of the shallow ocean, almost all iron in seawater is immediately oxidized and removed from solution. The main pulse of banded iron formation precipitation is thought to record a time when oxygen levels in the atmosphere were increasing as a result of photosynthesis, and the reduced iron that had been held in solution was oxidized and deposited as iron-rich sediment.

It was not until virtually all of the iron in the oceans had combined with the newly available oxygen that any free oxygen could begin to accumulate in the atmosphere.

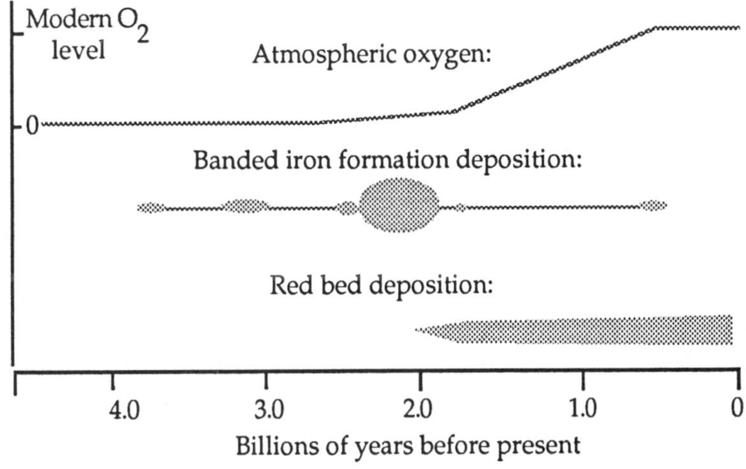

a) <u>Red beds</u> are continental sedimentary deposits containing red-colored oxidized iron. Study Figure 6-1 and describe the relationship between the times of deposition of banded iron formations and red beds. Why do you think this relationship exists?

Rusting, decay and burning are all oxidation reactions. If atmospheric oxygen levels were much higher than present values, global forest fires would result.

If photosynthesis were to cease today, Earth's atmosphere would gradually return to a low-oxygen state and primitive anaerobic organisms would again dominate the biosphere.

Gaia hypothesis: view that Earth's near-surface environment has been largely controlled by biological activity -- that organisms have acted to minimize fluctuations in surface temperature and air and water chemistry

natural selection: evolution of organisms by the selective "survival of the fittest"

model: simplified representation of a system

b) Examine the iron formation and red bed specimens, then make sketches and brief descriptions of them.

The dramatic increase in atmospheric oxygen in the middle Precambrian paved the way for development of the diverse life forms on the modern Earth, but it was deadly to many of the organisms that had evolved in the CO_2-dominated atmosphere of the early Earth. Today, the descendants of some of these oxygen-shunning (anaerobic) organisms live in environments like swamps, the depths of stagnant water bodies and the stomachs of cows.

III. The Gaia hypothesis

In the preceding exercises, you have discovered that the elements phosphorous, nitrogen and oxygen, all critical to life, would not be available to organisms if they were not concentrated and constantly replenished by the biosphere itself. Many other chemical cycles, including those of water, carbon and sulfur, are also profoundly linked with Life on Earth.

Atmospheric scientist James Lovelock has proposed a new view of Earth that highlights the central role of the biosphere in mediating the chemistry of Earth's air, water and sediments. Lovelock maintains that life has not merely adapted to a hostile environment but has instead actively controlled the environment for its own benefit for billions of years. At the suggestion of novelist William Golding, Lovelock named his proposal the <u>Gaia hypothesis</u> after the Greek goddess of Earth.

Lovelock suggests that the Earth is, in some sense, a superorganism capable of regulating its own temperature and chemistry in much the same way that humans and other organisms maintain such internal balances. Lovelock cites evidence that Earth's surface temperature and atmospheric composition have remained remarkably constant for at least 2 billion years in spite of dramatic changes in the configurations of tectonic plates, geography of the oceans and even the energy radiated by the Sun.

Critics of the Gaia hypothesis argue that it is incompatible with Darwin's theory of evolution by <u>natural selection</u> -- that individual organisms acting in their own self-interest could not control Earth's environment on a global scale. These critics assert that the Gaian view requires some sort of central organizing entity. Lovelock has developed a simple <u>model</u> to illustrate how a Gaian system could be controlled by organisms concerned only with their own survival. This model is called Daisyworld.

A. Daisyworld

Daisyworld is a hypothetical Earth-like planet, the same size as Earth and orbiting the same distance from a star similar to Earth's Sun. Like our Sun, this star has grown progressively brighter through time, radiating more and more heat. Yet the surface temperature on Daisyworld has remained nearly constant for most of the planet's history. This is because the biosphere on Daisyworld, which consists only of dark-, light-, and gray-colored daisies, has acted to moderate the temperature. The daisies influence the surface temperature simply through their <u>albedo</u> or reflectivity. Dark daisies absorb most of the Sun's heat; light-colored daisies reflect much of it back to space. Gray daisies absorb about as much heat as they reflect. But how could the reflectivities of individual daisies affect the global temperature?

Early in the history of the planet, when the young Sun was still relatively cool (Fig. 6-2), dark daisies would be the fittest species, because clusters of them create local warm spots that favor the growth of more daisies. Soon the planet would be covered by dark daisies, and their collective effect would be to increase the global temperature above what it would have been in the absence of life (time A, Fig. 6-2).

When the dark daisies had established a comfortable temperature, gray and white daisies would begin to take advantage of the pleasant conditions. At first, gray daisies would do better than light ones because clusters of reflective light daisies wouldn't be able to keep local temperatures warm enough for survival (B). Eventually, the Sun's output would reach the point where unmoderated surface temperatures would exceed the maximum tolerable to daisies (C). At this point, light-colored daisies would begin to become the fittest species because clusters of them would create cool spots that would favor the growth of more daisies. As light-colored daisies spread, their collective effect would be to decrease the global temperature well below what it would have been in the absence of any life forms (D). In this way, individual daisies, without knowledge of or concern for the planet as a whole, would have acted to control the global environment.

Finally, the heat produced by the Sun would be so great that neither type of daisy would be able to moderate the temperature, and all species would die out (E).

albedo:
the percent of incoming heat or light energy reflected by an object

If you have sat on a hot black car seat in the summer or have been dazzled by the brightness of snow on a sunny day in winter, you already know that dark-colored objects tend to absorb heat and light while light-colored objects reflect them.

FIGURE 6-2.
TEMPERATURE AND POPULATION VS. TIME ON DAISYWORLD
In lower plot, shaded area indicates temperature range tolerable by daisies.

Reproduced from *The Ages of Gaia: A Biography of Our Living Earth,*, by James Lovelock, with permission of W.W. Norton and Company. Copyright 1988 by the Commonwealth Fund Book Program of Sloan-Kettering Memorial Cancer Center.

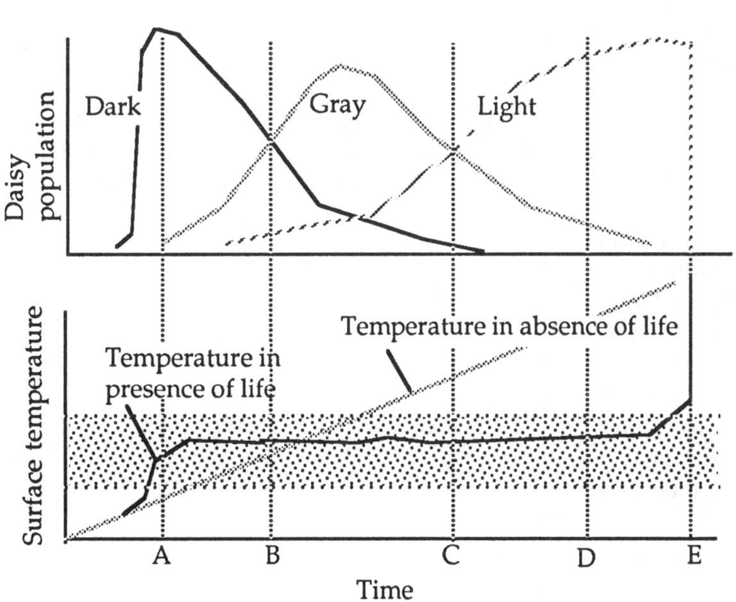

1. On the lower plot in Figure 6-2, mark and label the time interval over which temperatures would have been in the tolerable range in the *absence* of the daisies. Then mark and label the time interval over which this temperature range would have been sustained in the *presence* of the daisies. Explain the significance of the difference between these two time intervals.

2. On the upper plot in Figure 6-2, sketch a curve showing how the population of daisies (of any color) would change over time if the daisies had *no* effect on surface temperature. Explain the placement and shape of your curve.

3. One implication of Daisyworld is that life, if it is to survive for any length of time on a planet, cannot be sparse; it must exist in abundance. Explain this statement.

The seeds for the Gaia hypothesis came from James Lovelock's work with NASA in the 1960s when he worked on a project to determine whether there was life on Mars....

4. Write a short critical analysis of the Daisyworld model. In what ways is it a reasonable representation of the role of life on Earth? In what ways is it an oversimplification? Hint: Consider the nitrogen, phosphorous, or carbon cycles.

...Lovelock recognized that determining the composition of the Martian atmosphere was sufficient to establish that the planet is lifeless; there are none of the telltale gases indicative of life.

ozone layer:
zone in Earth's stratosphere between about 20 and 40 km where ozone concentra-tions are at their highest

Ozone reacts readily with a variety of compounds, especially those containing chlorine. Human production of chlorine aerosols threatens to destroy the fragile ozone layer that acts as Earth's natural sunscreen.

B. Earthly examples
Biological intervention in global conditions can be visualized for the idealized case of Daisyworld, but does the Gaian model really apply to systems as complex as those on Earth? Although it may never be possible to chart all of the strands in the web that links the biosphere with other Earth systems, biological control of the environment has been documented for parts of this web. Among the most interesting examples are two atmospheric gases that are both critical for and created by life.

Ozone (trioxygen, O_3) is a minor component of Earth's atmosphere by volume but vital to life on Earth because it is the only molecule that absorbs cell-damaging wavelengths of ultraviolet radiation from the Sun. Yet ozone would not be present in the atmosphere if not for the existence of diatomic oxygen (O_2) produced by green plants. The ozone layer between altitudes of about 20 and 40 km is produced by the conversion of O_2 to O_3, with solar ultraviolet radiation providing the energy for this reaction.

Dimethylsulfide (DMS) is a trace gas produced mainly by marine algae. In high concentrations, DMS acts to nucleate clouds, which regulate incoming solar radiation and ocean temperatures. This in turn facilitates continued growth of algal colonies, the basis of the oceanic food chain.

1. Lovelock argues that examples like ozone and dimethylsufide show that the Earth can be considered a superorganism with internal organs and control systems analogous to those in an individual animal or plant. Do you agree? Why or why not?

2. Can you think of other instances in which a species or group of species modifies the environment in a way that results in collective benefit? How do you think such feedback systems arise over time?

IV. Synthesis

A. Preview review
Return to the Lab Preview and complete any questions you were unable to answer before.

B. You are what you eat
Find a book containing nutritional tables and determine the best food sources of the elements iron, zinc, magnesium and phosphorous. Explain why each of these nutrients is important to human health. Then speculate about how the elements came to be concentrated in these foods.

C. We are what we discard
There have been many proposals about what sets human beings apart from other members of the animal kingdom -- for example that we alone have language or consciences or technology. Perhaps the most distinguishing characteristic of our species, however, is that we produce waste materials which have no part in global chemical recycling systems. We alone fail to "close the loop", and therefore cause imbalances in input and output fluxes of geochemical reservoirs. Comment on this assertion. Is this an inherent human trait or one that appears only in certain cultures?

D. Old Mother Earth
Write a brief essay summarizing your views on the Gaia hypothesis. In light of all that you have learned about interactions among the biosphere, atmosphere, hydrosphere and solid Earth, do you think it is a useful way to think about life on Earth? Is it a testable scientific hypothesis or nothing more than a metaphor? Justify your stance.

E. New Mother Earth
Biosphere 2, an attempt to create a completely closed, self-sustaining miniature ecosystem in the Arizona desert, failed during a 2-year trial in 1991-92 because atmospheric oxygen levels in the sealed environment fell dangerously low (14%) for the eight human inhabitants. The disappearing oxygen was eventually traced to the organic-rich soil, where bacteria were using it to convert organic carbon to CO_2. *Biosphere 2* illustrates the difficulty of replicating Earth's balanced chemical budgets. (*Biosphere 1* is the Earth itself).

Do you think efforts like *Biosphere 2* will ever succeed? In what ways is this sort of created environment fundamentally different from natural Earth systems? Do you think it will be possible to "transplant" life to planets like Mars? Justify your answers with specific reasons or examples. (See article by Severinghaus et al. in references on next page).

SUGGESTED REFERENCES AND RESOURCES

Biogeochemical cycles:

National Research Council, 1986. *Global Change in the Geosphere-Biosphere: Initial Priorities for the International Geosphere-Biosphere Program.* Washington DC: National Academy Press, 91 pp.
> Includes concise summaries of natural biogeochemical cycles, records of past atmospheric change, and projected effects of human perturbations of these systems.

Schlesinger, W. H., 1991. *Biogeochemistry: An Analysis of Global Change.* San Diego: Academic Press. 443 pp.
> One of the first textbooks to provide an overview of the field of biogeochemistry. Pitched at the advanced undergraduate/graduate level, the book assumes readers have a working knowledge of chemistry.

Atmospheric composition and evolution:

Wayne, R., P., 1991. *Chemistry of Atmospheres* (second ed.). Oxford: Oxford Science Publications. 430 pp.
> A readable textbook that emphasizes the processes that link the atmosphere with the biosphere, hydrosphere, and solid Earth. Also includes sections on the atmospheres of other planets.

Gaia:

Lovelock, J., 1988. *Ages of Gaia.* New York: W. W. Norton, 252 pp.
Lovelock, J., 1989. Geophysiology: The science of Gaia. *Reviews of Geophysics,* v. 27, p. 215-222.
Lovelock, J., 1990. Hands up for the Gaia hypothesis. *Nature,* v. 344, p. 100-102.
> Lovelock's own recent words on Gaia. The latter two articles appear in mainstream science journals and address some of the issues raised by skeptics.

Schneider, S., and Boston, P., eds., 1992. *Scientists on Gaia.* Cambridge, MA: MIT Press, 431 pp.
> Proceedings of an American Geophysical Union-sponsored conference on Gaia that brought together geologists, biologists, philosophers and historians of science.

Williams, G., 1992. Gaia, Nature worship and biocentric fallacies. *The Quarterly Review of Biology,* v. 67, p. 479-486.
> A strongly worded attack on the Gaian point of view. Provides good material for lively debate.

Maxis Corp., 1992. *SIMEARTH* software.
> A computer game based loosely on the Gaia hypothesis. Players can vary a range of environmental, social and political parameters and see their effects on an Earth-like planet over time. Maxis 1042 Country Club Dr., Suite C, Maraga, CA 94556. Phone: 415/376-6434. About $60.

Biosphere 2:

Severinghaus, J., Broecker, W., Dempster, W., MacCallum, T., and Wahlen, W.,1994. Oxygen loss in Biosphere 2. *EOS, Transactions, Am. Geophysical Union,* v. 75, no. 3.
> A fascinating analysis of the carbon and oxygen budgets in Biosphere 2.

EARTH RESOURCES

What are the geologic origins of common Earth resources?
How has resource consumption changed throughout history?
Why is there disagreement about projections of resource exhaustion?

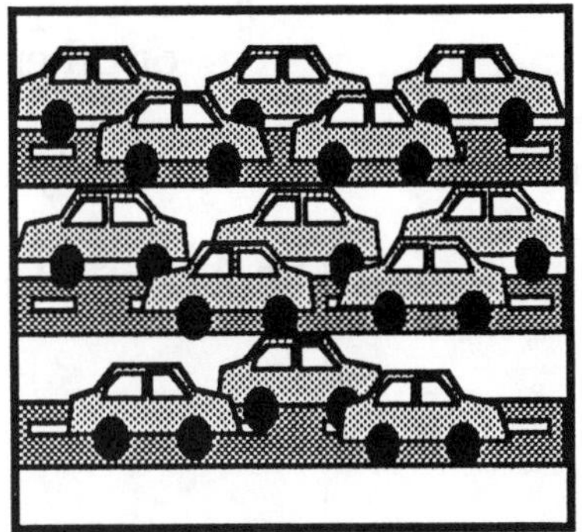

The importance of Earth resources in shaping civilizations is reflected in the names ascribed to stages of human history: the Stone Age, the Bronze Age, the Iron Age, and the Atomic (Uranium) Age.   Until the rise of modern civilization, however, consumption of Earth resources was negligible.

Today, in contrast, it is estimated that every year each U.S. citizen accounts for the consumption of 1300 pounds of steel and iron, 65 pounds of aluminum, 25 pounds of copper, 15 pounds of manganese, 15 pounds of lead, 15 pounds of zinc, and 35 pounds of other metals. Energy resources associated with use of these metals includes 8000 pounds of oil, 4700 pounds of natural gas, 5150 pounds of coal, and 1/10 of a pound of uranium (Youngquist, 1990. *Mineral Resources and the Destinies of Nations*).

In Laboratory 2 (*Earth in Time*) you discovered the power of the principle of Uniformitarianism -- the idea that "the present is the key to the past". Although this is true for most geologic processes, it is certainly not true for the use of Earth resources. In the last century, technological advances and population growth have led to extraction of Earth resources at rates that are orders of magnitude faster than the geological processes that originally concentrated the resources. The consequences of this imbalance in consumption and replenishment rates have yet to be realized.

OBJECTIVES

- To examine models of human resource consumption
- To explore the uncertainties associated with predicting future resource availability

OUTLINE

I. Metal resources
    A. Aluminum: A case study
    B. Lead: A case study

II. Energy resources
    A. Petroleum
    B. Other fossil fuels
    C. Alternative energy sources

<u>LAB PREVIEW</u>   Before coming to lab,
- Answer as many of the following questions as you can.
- Read Chapter 17: "Resources from the Earth" in *The Blue Planet* or other readings assigned by your instructor

How much do you know already about Earth resources?

What raw Earth resources do you use in everyday life?

Where do these resources come from, and how abundant are they?

How has the population of the Earth changed over time, and how has that change affected resource consumption? What other factors govern rates of resource use?

How are predictions about resource depletion made and what are the major sources of uncertainty in these predictions?

How much longer can remaining supplies of petroleum be expected to last?

What is meant by a renewable resource? Do fossil fuels fit this definition?

What products of fossil fuel burning can have negative environmental effects?

What problems are associated with other energy sources?

I. Metal resources

Use of Earth resources is often described as "consumption" but is perhaps better viewed as "redistribution", or reorganization of matter on Earth. Like Earth's water, the planet's mineral resources are a legacy of its earliest history. Almost nothing enters or leaves the system, though tectonic and surficial processes -- and now human activities -- are constantly changing the form and distribution of mineral resources. When humans use mineral resources, matter is neither created nor destroyed, though energy is consumed in changing the raw materials to usable forms.

Mineral resources are not uniformly distributed on or within the Earth. Nations may be rich in some resources (for example, the U.S. holds 58% of the world reserves of molybdenum), but lack adequate supplies of others (the US must import chromium and petroleum). The non-uniform distribution of Earth materials leads to trade in good times and war in bad times. Like all commodities, mineral resources are valued according to their availability.

A. Aluminum: A case study

Historical use of aluminum is an excellent illustration of the non-uniformitarian nature of resource consumption. Before 1885, no simple chemical process was known for refining aluminum. Pure aluminum metal was extremely rare and therefore very valuable, and worldwide consumption was negligible. In 1852, aluminum cost $545 per pound (in 1852 dollars!).

In 1885, Charles Hall, an undergraduate chemistry major at Oberlin College (Ohio), began experimenting with aluminum refinement and developed an inexpensive method for extracting aluminum from ore. He established the Aluminum Company of America (Alcoa) and became a wealthy man. A single technological breakthrough made a once scarce commodity widely available.

In the exercises below, you explore the redistribution of one mineral resource by tracking the history of aluminum used in soda cans.

1. Aluminum in Earth's crust

Aluminum (Al) is the third most abundant element in the crust (Table 7-1) and is present in many common silicate minerals (feldspars, micas, garnet). Aluminum atoms are bound so strongly with oxygen atoms in these minerals, however, that a tremendous amount of energy is required to separate them.

Examine the specimens of common aluminum-bearing minerals. You have seen most of these minerals before; in what rock types do they most commonly occur?

Napoleon, among the world's richest men in his time, had a prized set of dinnerware made of pure aluminum, then one of the rarest metals known.

ELEMENT	WT. %	ELEMENT	WT. %
Oxygen (O)	46.6	Carbon (C)	0.020
Silicon (Si)	27.72	Manganese (Mn)	0.095
Aluminum (Al)	**8.13**	Sulfur (S)	0.026
Iron (Fe)	5.00	Barium (Ba)	0.043
Calcium (Ca)	3.63	Chlorine (Cl)	0.013
Sodium (Na)	2.83	Chromium (Cr)	0.010
Potassium (K)	2.59	Fluorine (F)	0.063
Magnesium (Mg)	2.09	Zirconium (Zr)	0.016
Titanium (Ti)	0.44	Nickel (Ni)	0.008
Hydrogen (H)	0.14	**Lead (Pb)**	**0.001**
Phosphorous (P)	0.11	Others	0.011

Source: Mason B., 1966, *Principles of Geochemistry*, New York: Wiley.

ore:
economically viable
source of a
metallic element

2. Aluminum ores

Although aluminum is very abundant on Earth, extracting it from silicate minerals like those you just examined requires too much energy to be practical. Instead, aluminum is extracted from <u>ore</u> minerals in which the element is bound less tightly. The distribution of these ore minerals is controlled by climatic factors; the vast majority of the world's aluminum supply comes from the tropics or areas that once had tropical conditions in the geologic past.

a) What climatic conditions distinguish the tropics from other regions?

The high aluminum levels in tropical soils can become toxic to plants if the soil pH falls below about 4.5 as a result of acid rain (see end of this laboratory).

b) The climatic conditions you identified break down aluminum-bearing silicate minerals, forming oxides and hydroxides of aluminum such as the mineral gibbsite [Al(OH)$_3$]. Soil and rock made up of these new aluminum minerals are called <u>laterite</u> and <u>bauxite</u>, respectively. Examine the specimens of these and other metal ores and briefly describe their characteristics (color, hardness, density, crystal form, etc.).

The U.S. lacks major aluminum deposits and currently imports 97% of its aluminum ore.

c) Price of aluminum ore

Like other commodities, the price of aluminum ore fluctuates daily on world financial markets. What is the current price of one kilogram of aluminum? Check the financial pages of the newspaper.

How much did the US pay to other nations for the 6,214,000 metric tons of Al it consumed in 1988? Use today's price for Al.

Aluminum refining is a major industry in Iceland, where geothermal energy supplies abundant, inexpensive electrical power.

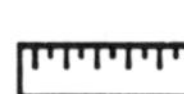

3. Energy costs of refining aluminum

The aluminum in laterite and bauxite must still be separated from oxygen and hydrogen, and then purified. Although these bonds are individually quite weak, a large amount of electrical energy is required to obtain any usable amount of aluminum . For example, processing 1 kg of aluminum from ore consumes 15.55 kilowatt hours of electricity. Aluminum refining is profitable only where energy is inexpensive.

a) The most familiar use of aluminum is for beverage containers. The aluminum in soda cans was extracted, processed, and shipped simply to transport a drink from one place to another. What are more efficient ways?

b) How much does one aluminum can weigh?

At \$0.08/kilowatt hour, what is the cost of refining the aluminum in one can?

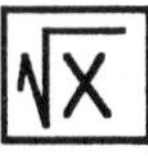

c) Using the approximation of 1 can/day per person, calculate the cost of the energy used in the refinement of aluminum cans consumed by your college/university community in one year. (If an actual estimate of aluminum use for your campus is available, use that figure in your calculations).

d) The US consumed 6,214,000 metric tons (1 metric ton = 1000 kg) of aluminum in 1988. At the rate cited above, what was the cost of the energy required to refine this aluminum? Show your work.

e) It takes only 6% as much energy to recycle aluminum as to produce from raw ore. If all the aluminum in question d was recycled aluminum, what would have been the cost of the energy used to produce it?

At present, only about half of all aluminum cans are recycled in the US.

B. Lead: A case study
We have considered the economics of aluminum, an element that is abundant but requires large amounts of energy -- a scarcer resource -- in its processing. We now examine another important metal, lead, that is not at all abundant on Earth.

As indicated in Table 7-1, lead (Pb) exists in the Earth at a concentration of 0.0013%, or 13 parts per million (ppm). In other words, for any million units of weight of the crust, 13 of those units will be lead. Though rare, lead is a strategic commodity, necessary in the transportation, communications and electronics industries, principally for electrical power storage (batteries).

1. Lead in Earth's crust
a) As you learned in Laboratory 3 (*The Solid Earth*), the composition of the continental crust is close to that of granite. Consider the large piece of granite in the lab to be representative of the continental crust. What is the mass of the rock, in grams?

b) How much lead does this piece of granite contain, assuming it is an "average" piece of continental crust? Show your calculations.

c) What is the typical density of granite and continental crust? (See Table 3-1).

Using this density, determine the mass (in grams) of a cube of continental crust one meter (= 100 cm) on an edge. Show your work.

d) Given the average crustal concentration of 13 ppm, how much lead (in grams) is contained in that cubic meter of crust?

e) A 12-volt car battery typically contains about 14 kg of lead. What volume of average continental crust (in m^3) would have to be processed to extract the lead necessary for one car battery?

The US government maintains stockpiles of lead for strategic use in case supplies were to be cut off.

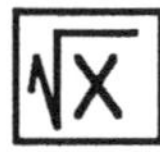

Until the mid-1970s, lead was widely used as a gasoline additive. Each car of that era emitted some 225 kg of lead in its lifetime.

Lead from car exhaust entered the air and soil where it was taken up by plants and animals including humans.

Today, almost all lead used in storage batteries is recycled. Recycling rather than discarding lead became a priority after its toxic nature was recognized in the 1970s.

2. Lead ores

Obviously, the average crustal abundance does not accurately characterize the availability of lead. It would be impossible to process such volumes of rock to extract the lead consumed by society.

In practice, lead is extracted from the mineral <u>galena</u>, or lead sulfide. Galena is made of atoms of lead bonded to atoms of sulfur (S) (Figure 7-1), and its chemical formula is PbS . Deposits of galena form only under specific geologic conditions, typically where hot, briny groundwater has moved through rock and left behind precipitates. Economic geologists attempt to locate such deposits, evaluate their profit potential, and eventually oversee extraction of the deposits in mining operations.

Like most mineral resources, lead ores are not uniformly distributed around the world. The US, Canada and Australia hold half of all global lead reserves.

There is not, however, an unlimited supply of galena in the crust. Because lead is a strategic metal, there is great interest in the future availability of lead as a resource. Several governmental and private agencies keep a close watch on the supplies of extractable lead remaining in the Earth, and what countries control those supplies.

FIGURE 7-1.
ATOMIC STRUCTURE OF THE MINERAL GALENA (PbS)
Large white spheres represent sulfur atoms; small gray spheres lead.
Rods depict bonds between atoms.

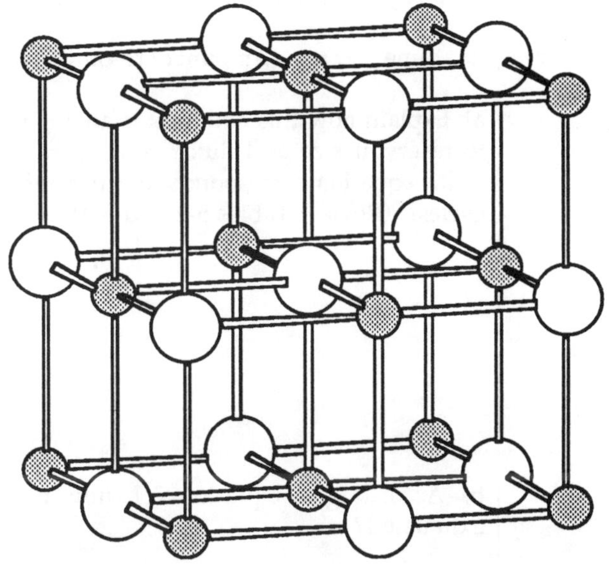

a) Examine the specimens of galena and other ore minerals. Make a sketch showing the crystal form of galena. Does this seem consistent with its atomic structure (Fig. 7-1)? Explain.

What property distinguishes galena from the other minerals? (Close your eyes).

b) Why do you think that galena has this property? Refer to a periodic table of the elements.

Like many other Earth resources, lead ore is consumed by industrial society at rates that far outstrip the rates at which it accumulated by natural geologic processes. Before about 1970, however, there was little public awareness about the rapid depletion of Earth resources; metals, fuels and water all seemed inexhaustible commodities. Then in 1972, a ground-breaking book, *The Limits to Growth*, showed not only that most Earth resources are exhaustible, but that many could actually be used up within a lifetime.

In the exercises below, you will recreate some of the calculations done in *The Limits to Growth*, using the example of global lead consumption. Instructions for carrying out the exercises on a spreadsheet appear in Appendix I (p. 174).

3. Future lead use at static consumption rates

You can make estimates of the future availability of lead using several different scenarios. First, use the facts that in 1989 worldwide reserves of lead were estimated at 120,000,000 tons and the worldwide consumption of lead was 6,000,000 tons/year. Next, assume that consumption occurs at a <u>static</u> rate, i.e., that humankind will consume the same amount of lead (6,000,000 tons) every year. By this model, the time to exhaustion is simply:

static:
fixed or unchanging

$$\text{Time} \quad = \quad \text{(Amount of resource available)/(Annual consumption)} \qquad (7\text{-}1)$$

a) Explain how this equation is analogous to the equation relating residence time to reservoir size and flux rate (Laboratory 5, *Earth's Aura*, eq. *5-1*). Which term in the equation corresponds to flux rate? How is it different from fluxes in natural cycles? (Review Tables 5-2 and 5-4).

b) According to equation *7-1*, how many years remain until lead supplies are exhausted?

By this estimate, how old will you be when lead supplies are exhausted?

c) What is the important -- and probably inaccurate -- assumption behind this model?

4. Future lead use at increasing consumption rates

The main assumption in the static model is that the consumption of lead is constant. Lead is consumed by humans, and as the human population increases, so does the consumption of lead. Figure 7-2 shows how world population has changed since 8000 BC. Note that over the past several centuries, population growth is not <u>linear</u>, but <u>exponential</u>. That is, world population does not increase by a constant number of people each year but instead at a rate of about 2%/year. This is analogous to a simple interest rate problem in which the "principal" is the population in about 1650 and the "interest" accumulates at 2%/year.

Mathematically, we can describe exponential population growth with the equation:

$$P = P_0 e^{ct}, \qquad\qquad (7\text{-}2)$$

where P is the population at time t, c is a constant that depends on the growth rate, P_0 is the population at some starting time ("time 0"), and e is the base of the natural logarithms, equal to about 2.7183.

The same equation can be used to model the progressive depletion of resources at increasing rates of consumption. Because the number of resource consumers grows exponentially, so does resource consumption.

**FIGURE 7-2.
WORLD
POPULATION
SINCE 8000 B.C.**
Source: Population
Reference Bureau.

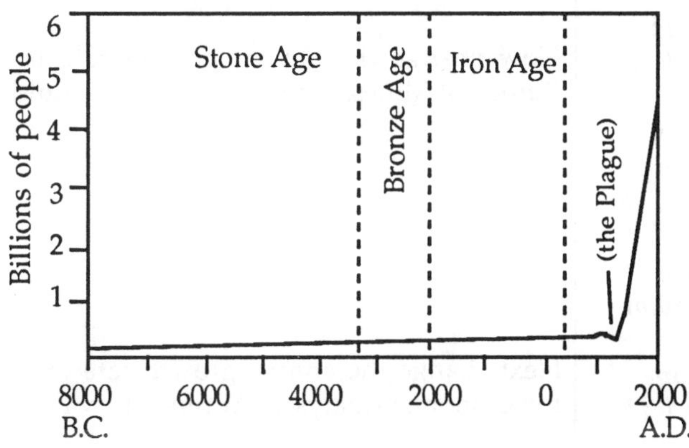

a) Assume that since 1989, the consumption of worldwide lead resources has increased 3%/year (i.e. each year 3% more lead is consumed than in the previous year). How much lead would have been consumed in 1990, compared to the 6,000,000 tons in 1989? Show your calculations.

b) Using the same approach, project how much lead would be consumed in 1999. (To determine consumption for a given year, simply multiply the previous year's consumption by 1.03 Repeat this for each year between 1990 and 1999). Make a table or plot showing your projections of lead consumption.

c) The time to resource depletion can be found by the equation:

$$\text{Time} = \frac{ln[\,(\text{Growth rate} \times \text{yrs. to exhaustion in static model}) +1\,]}{\text{Growth rate}} \quad (7\text{-}3)$$

where ln is the natural logarithm.

For example, if an Earth resource would be exhausted in 20 years assuming static rate of consumption, but the consumption actually increases by 2%/year, the years to depletion would be:

$$\text{Time to depletion} = \frac{ln\,[(0.02/\text{yr.} \times 20\,\text{yrs.})+1]}{0.02/\text{yr}} = 17\ \text{years}$$

According to this exponential model, when will the supply of lead be exhausted? (Use eq. 7-3, and start with the 1989 data given previously).

Some historians believe that lead poisoning may have contributed to the fall of the Roman Empire. Wealthy Romans used lead for dinnerware and water pipes. The irrational behavior of some Roman leaders may have been caused by lead-induced brain damage.

Now change the annual growth rate to 0.9%, the rate predicted by the U.S. Bureau of Mines. At this rate, when would lead reserves be exhausted?

Next change the annual growth rate to 4.8%, the rate predicted by the Institute for Economic Analysis. At this rate, when would lead reserves be exhausted?

d) Why do you think there is such a wide range in the estimates of the rate at which a resource will be used in the future? What factors would tend to *decrease* the rate at which a raw resource is consumed? What technological changes made since 1970 have affected lead consumption?

5. Estimates of remaining reserves
The mathematical models you have used above are based on several critical assumptions. Among the most important are assumptions about remaining lead reserves. How can estimates of remaining reserves-- i.e. still undiscovered lead -- be made?

Predictions of reserves are based on knowledge of the geologic conditions that favor accumulation of particular resources. These predictions are often wrong by factors of two or three because it is simply not possible to observe what remains within the Earth. How much do changes in estimated reserves change predictions about resource depletion?

a) Assume that the amount of remaining lead in 1989 was *underestimated* by a factor of 5 (i.e. that there were actually 600,000,000 tons of extractable lead remaining). Based on this estimate, when would lead supplies be exhausted, assuming a *static* rate of consumption of 6,000,000 tons/year? (Use eq. *7-1*).

b) Now use the revised estimate of remaining reserves in the exponential model with a 4.8% annual growth rate and determine when the last consumable lead would be extracted . (Use eq. *7-3*).

c) This projection assumes that geologists have underestimated the amount of remaining consumable lead by a full factor of five. But how many years does this add to the predicted time to depletion for a growth rate of 4.8%?

By this estimate, how old will you be when the last extractable lead is mined?

Explain why increased reserve estimates have relatively little effect on the predicted time to depletion for exponential growth models.

II. Energy resources

Most of the energy that we exploit originates as <u>hydrocarbons</u>. As their name implies, hydrocarbons such as petroleum are formed principally of hydrogen and carbon. Together with oxygen, these elements are the main components of the organic, or "living" portion of the Earth, including our own bodies (Lab 6).

Energy resources including petroleum, coal and natural gas are termed "fossil fuels" because the hydrogen and carbon that form them were derived from living organisms tens or hundreds of millions of years ago; the fuels represent "fossils" of the original organisms. Fossil fuels are <u>non-renewable</u> energy resources because the time required to form them from organic matter is, on a human time scale, essentially infinite.

A. Petroleum

Imagine driving down an unfamiliar road. It is late, and you look at your fuel gauge. Because the way is unfamiliar, you don't know how far it is to the next gas station. Which of the gauge readings would make you concerned about finding more gas?

Of the four gas gauges, the one in the lower left represents the amount of petroleum currently remaining from the original US resources. The fuel gauge in the upper right represents the approximate remaining petroleum in the world.

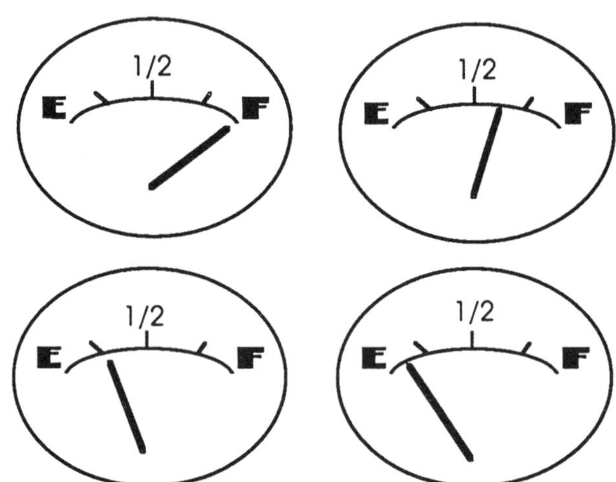

Few discoveries have changed the course of human history more than the discovery of oil. Before the first successful oil well was drilled in 1859, world oil production was only a few barrels a day, all of it collected from "seeps" where oil came naturally to the surface. Today, world production exceeds 50,000,000 barrels daily (one barrel = 42 gallons). One needs only to envision life as it was 150 years ago to imagine what life would be like without petroleum.

The discovery of oil and the ability to tap it in large volumes allowed unprecedented breakthroughs in transportation, agriculture, heating, cooling, lighting, and extraction of other Earth resources. It transformed humankind into a force capable of causing geologic change at rates far greater than natural processes.

What is the future of the resource that has shaped the modern age? Can ever-increasing levels of energy consumption be sustained? In this section, you will explore the history and future of petroleum consumption at national and global levels.

Do you think the prices of these products accurately reflect their value? Consider which are locally produced, which are renewable, which require the most processing.

1. What is a barrel worth?

Given our society's heavy dependence on petroleum, its current value on world markets is unnaturally low when compared with other products we consume in our daily lives. In 1994, crude oil cost about $18 per barrel (42 gallons). For comparison, estimate the "per barrel" cost of the following products.

	per gallon	per barrel		per gallon	per barrel
Gasoline			Bottled water		
Milk			Beer		
Orange juice			Wine		
Brand-name soda			Perfume		

2. Historical petroleum production

The petroleum we use accumulated in sedimentary rocks over the past half billion years, but it is quite possible that this 500-million-year legacy will be exhausted within your lifetime. Table 7-2 summarizes global production and consumption of petroleum since 1860.

a) What major political and economic trends and events can you recognize in the patterns of petroleum production?

TABLE 7-2.
GLOBAL
PETROLEUM
PRODUCTION
SINCE 1860

Decade	Global Production	Cumulative global production
	Millions of barrels	
1860	29	29
1870	119	148
1880	409	557
1890	1033	1590
1900	2186	3776
Total Global Consumption 1860-1909 = 3.776 billion barrels		
1910	4282	8058
1920	10,567	18,625
1930	16,617	35,242
1940	26,410	61,652
Total Global Consumption 1910-1949 = 57.876 billion barrels		
1950	53,689	115,341
1960	109,297	224,638
1970	200,388	425,026
1980	203,754	628,780
Total Global Consumption 1950-1989 = 567.128 billion barrels		

Sources:
Campbell, C.J., 1991. *The Golden Century of Oil*, 1950-2050. Boston: Kluwer Academic Publishers.
Karlsson, S., 1986. *Oil and the World Order*. New York: Berg Publishers, Ltd.

 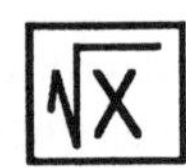

b) What percent of total global petroleum consumption through 1989 occurred in the five decades between 1860 and 1909? Use the multi-decade consumption values given in Table 7-2.

What percent of total global consumption through 1989 occurred in the four decades between 1910 and 1949?

What percent of total global consumption through 1989 occurred in the four decades between 1950 and 1989?

c) Describe the overall trend in petroleum consumption indicated by your calculations. How can you account for this pattern of consumption?

3. Future petroleum production
The obvious question is how much longer petroleum resources can last. One of the earliest attempts to predict future petroleum production was made in the mid-1950s by geologist M. King Hubbert for the American Petroleum Institute. Hubbert used the "integral technique of prediction", which suggests that production rises from the time of resource discovery, through a maximum, then declines again to zero production. In a sense, this is an inverted form of the uniformitarian principle; *future* resource availability is predicted from the *past* production.

a) On the graph provided in Figure 7-3, plot decadal global petroleum production since 1860 using the data in Table 7-2 . Note that the vertical axis is in *billions* of barrels (1 billion = 1000 million).

b) Describe how the rate of production changed over the period from 1960 to 1989. Given this trend, when do you think petroleum production will reach its peak?

c) Project your production curve into the future by assuming that production in coming decades will be the mirror image of that in the past --i.e., that the curve will be symmetric about its maximum. On the timelines below the chart, shade in areas corresponding to the lifetimes of your grandparents, your parents and yourself. (You can approximate the times by assuming 30 years between generations and lifespans of 75 years).

d) According to your plot, during what decade will global production fall to the level of 1900? How old will you be then?

e) Compare your predictive curve with Hubbert's original curve at the end of this laboratory (p. 143). Are there substantive differences?

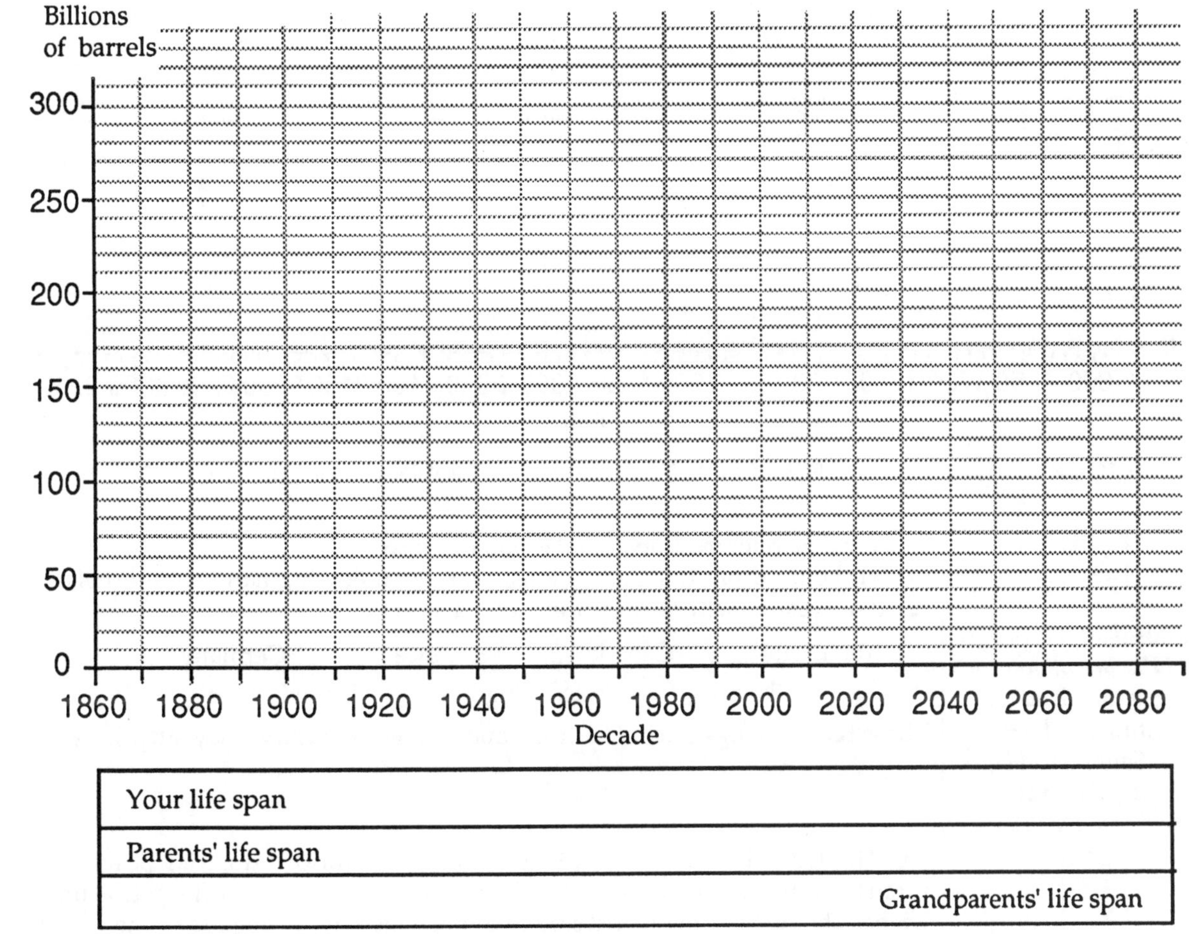

FIGURE 7-3. GLOBAL PETROLEUM PRODUCTION BY DECADE (HISTORICAL AND PROJECTED)

f) Your predictive curve illustrates the brevity of the Petroleum Age -- a mere instant of time in the 4.5 billion year history of the Earth. Even in the context of human history, the Petroleum Age is a fleeting period during which we are lucky to live. On the graph provided in Figure 7-4, plot global petroleum production as before, but this time with a slightly wider "window of time" in human history. (If you are using a spreadsheet, simply rescale the time axis on your plot).

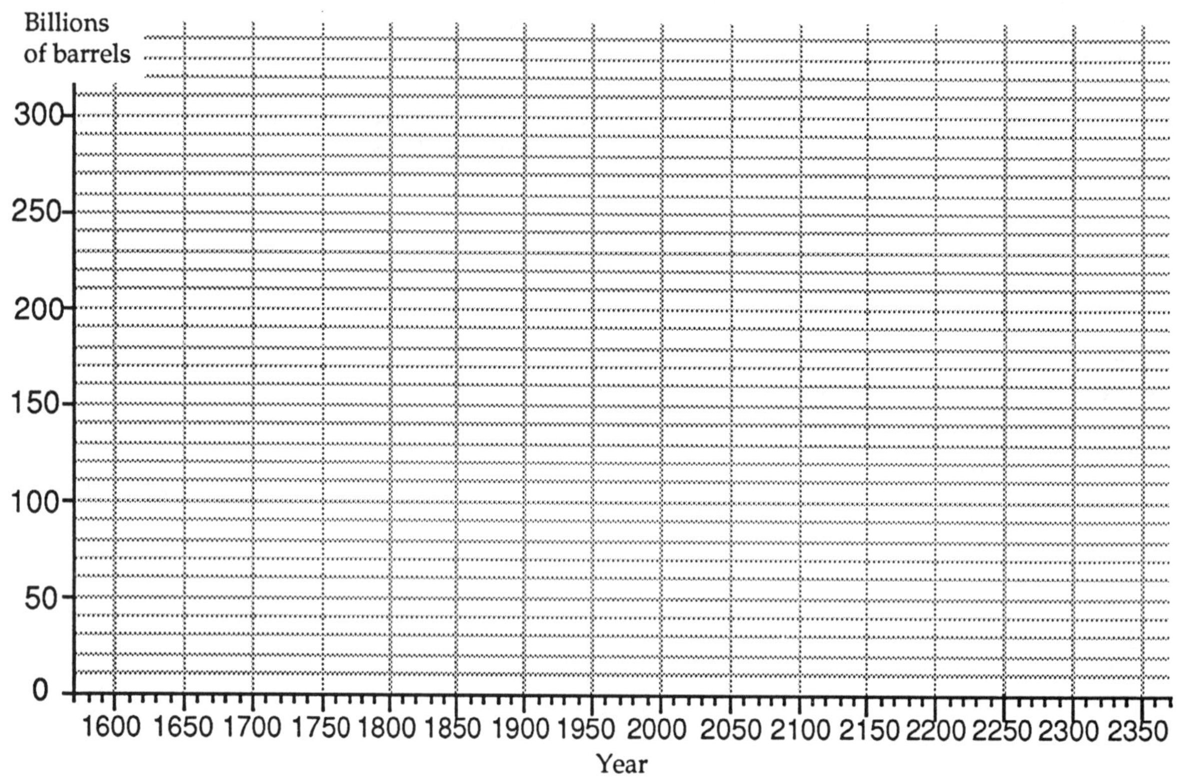

FIGURE 7-4. GLOBAL PETROLEUM PRODUCTION SHOWN OVER EIGHT CENTURIES

Almost half of the world's petroleum reserves are held by six countries: Iran, Iraq, Saudi Arabia, Kuwait, Neutral Zone, Abu Dhabi.

For the past 40 years estimates of global oil reserves have been essentially constant at about 1.6 $\times 10^{12}$ barrels.

4. Global petroleum reserves and economic power

a) On the basis of your perceptions, rank the following nations in terms of their relative wealth, with 1 indicating the richest.

Abu Dhabi___   Canada___   France___   Germany___

Japan___   United Kingdom___   United States___

The actual rankings (included at the end of this laboratory) may surprise you.

b) The United States attained its status as a world industrial power in part through consumption of disproportionate amounts of the world's petroleum. The US has about 6% of the world's population. Guess what fraction of the world's oil has been consumed by the US in recent decades. Compare this to the actual figure at the end of the laboratory. What is your reaction to this statistic?

At a time when petroleum prices are low, it is difficult to believe that this resource will ever be exhausted. Projections of the time remaining before global petroleum reserves are depleted differ, but there is no disagreement about the fact that the Petroleum Age will ultimately come to an end. What then?

Air quality in 19th-century London was deplorable. The legendary London fogs consisted mostly of coal smoke.

Carboniferous period: interval of geologic time 360-285 million yrs before present. A time of global coal formation. In North America, the period is subdivided into the Mississippian and Pennsylvanian periods.

The proverbial coals of Newcastle -- and most other British coal deposits -- are of Carboniferous age.

lignite: low-grade coal
bituminous coal: medium-grade coal
anthracite: high-grade coal

B. Other fossil fuels

The prospect of petroleum shortages has made other fossil fuels, particularly coal, seem attractive alternatives. Although coal was the fuel of the industrial revolution, it was largely supplanted by petroleum by the end of the nineteenth century. Coal burns less efficiently than petroleum and cannot be used in place of gasoline in internal combustion engines. Its principal advantage is its abundance

Coal is found on every continent; North America has reserves that could last hundreds of years. Coal deposits in Montana and Wyoming alone hold the energy equivalent of all the petroleum in Saudi Arabia. The ready availability of coal, however, needs to be weighed against the long-term environmental effects of its use. In Laboratory 5 (*Earth's Aura*) you discovered that the use of fossil fuels represents a significant human intervention into Earth's carbon cycle, and you considered the climatic implications of returning the long-stored carbon in fossil fuels to the atmosphere. The by-products of coal-burning pose additional threats to the environment.

In the following exercises, you explore the origin of coal and the benefits and problems associated with its use by reference to the state of Ohio, a significant coal-producer and the US state that consumes more coal than any other.

1. Origin of coal
a) Figure 7-5 shows the distribution of coal-bearing rocks in Ohio. What is the range in ages of coal-bearing rocks in Ohio? (see geologic time scale, Table 2-4).

The Pennsylvanian Period is part of a longer interval called the <u>Carboniferous</u> Period because it was a time when luxuriant forests drew large amounts of carbon from the atmosphere. Much of this vegetation escaped decay because it was preserved in swampy low-oxygen environments, and it was eventually transformed to coal. During this time, climatic conditions in Ohio were tropical, and plant life was lush and abundant.

b) These plants, like all organic matter, were made up primarily of the elements carbon (C), hydrogen (H), and oxygen (O). Where do plants obtain each of these elements? (Review Lab 6, *The Organic Earth*).

c) When the plants died, some were buried in sediment, typically in swampy deltaic areas. Their hydrogen and oxygen were expelled as water, leaving carbon and small amounts of other elements, which we will consider later. The heat and pressure associated with burial gradually changed the organic carbon into coal.

Study the specimens of different grades of coal. How does the appearance of the coal change with increasing grade?

FIGURE 7-5.
DISTRIBUTION OF
COAL-BEARING
ROCKS IN OHIO

2. Coal burning: Advantages

Burning coal is simply a process of oxidation, combining carbon with oxygen by the chemical reaction:

$$C + O_2 \to CO_2.$$

This reaction is <u>exothermic</u>, meaning that it gives off heat. This heat energy is used for generation of electric power.

The energy derived from 1 ton of coal equals that from 182 gallons of petroleum. How much petroleum was saved in Ohio by burning 32 million tons of coal in 1989?

How much oil was saved in the US that year by burning 981 million tons of coal?

exothermic reaction: chemical reaction that gives off heat

3. Coal burning: Drawbacks

Your answer to the question above illustrates why coal seems an attractive source of energy. But the energy gained from burning coal carries a hidden cost.

Most low and medium grade coal is not pure carbon. One of the impurities in most coal is sulfur (chemical symbol S), derived from the organic matter that formed the coal. Sulfur can occur in coal as mineral impurities such as pyrite (FeS_2) or gypsum ($CaSO_4$) or as "native sulfur", that is, pure sulfur distributed throughout the coal. High-grade coal tends to have less sulfur than low-grade coal.

When sulfur-bearing coal is burned (oxidized), sulfur combines with oxygen to form sulfur dioxide by the reaction:

$$S + O_2 \to SO_2.$$

In a matter of hours, this sulfur dioxide is further oxidized to SO_3. This in turn combines with atmospheric water in the reaction:

$$H_2O + SO_3 \to H_2SO_4,$$

where H_2SO_4 is sulfuric acid, a strong acid.

Coal-derived sulfur dioxide is the main culprit in acid rain formation, but nitrogen oxides from car exhaust also combine with water vapor to form nitric acid.

a) Examine the specimens of sulfur-bearing minerals, then look again at the lower grade coal specimens. Can you see pyrite or gypsum crystals in the coal? Scratch the native sulfur. Can you smell its distinctive "rotten egg" odor?

Acid rain can harm plants both directly, by damaging foliage, and indirectly, by altering their ability to absorb nutrients.

Because aquatic plants are particularly sensitive to low pH values, acid rain can seriously perturb food chains in lakes and streams.

Limestone and marble monuments in Athens, Rome and other European cities are being degraded by acid rain, which readily dissolves calcium carbonate.

b) Ohio coal contains between 1 and 6% sulfur; most of it is in the medium- (1.1 - 3%) to high- (>3.0%) sulfur range. Assuming an average sulfur content of 3.0% by weight, how much sulfur has been added to the atmosphere by the 3 billion tons of Ohio coal combusted through 1987?

Sulfur constitutes 32.7% of sulfuric acid by weight. If all the sulfur injected into the atmosphere was converted to sulfuric acid (H_2SO_4), how much H_2SO_4 was added to the atmosphere by burning Ohio coal through 1987?

c) How widespread is the problem of acid rain? Shade in the areas on Figure 7-6 where precipitation has pH values lower than that of "pure" rain water (Review discussion on pp. 91-92 of Lab 5).

In Lab 5 you measured the pH of rain that fell in your area. What was the value you found? Is this comparable to the value indicated for your region in Figure 7-6?

What common solutions have acidity levels comparable to rain in your area?

If your local rain is unusually acidic, is there a sulfur dioxide (or nitrogen oxide) producing area that can be identified as the source?

d) The Ohio River valley is the main source region for atmospheric SO_2 in the US. How can you explain the high acidity levels northeast of this area (Fig 7-6)?

FIGURE 7-6.
AVERAGE pH VALUES OF PRECIPITATION IN THE US AND SOUTHERN CANADA
Ohio valley region marked by dot.

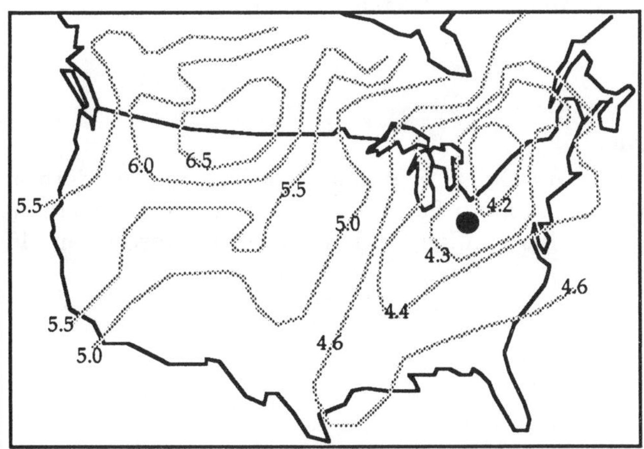

Source: Loucks, O., 1982. The concern for acidic deposition in the Great Lakes region. Chapter 2 in D'Itri, F., ed., *Acid Precipitation: Effects on Ecosystems.* Ann Arbor: Ann Arbor Science, p. 21-41. Reprinted with permission.

C. Alternative energy sources
The inevitable depletion of petroleum reserves together with growing concern about greenhouse gases and acid rain make the development of alternative energy sources a global priority.

In groups of two to three students, discuss and evaluate each of the following energy sources on the basis of the following criteria: 1) ability to meet demand for energy, 2) practicality for widespread use, 3) negative environmental impact, and 4) current status of technology. Refer to Chapters 17-18 in *The Blue Planet* or other readings recommended by your instructor.

Hydroelectric power	Geothermal power
Nuclear power	Wind power
Solar power	Tidal power

III. Synthesis

A. Preview review
Return to the Lab Preview and complete any questions you were unable to answer before.

B. Déjà vu
What type of recycling program exists on your campus or in your community? Contact the program coordinator or public relations manager and ask about the details of the operation. Which materials are accepted? What is the estimated participation rate? What incentives exist to encourage participation? Does the program make a profit? Which materials attract the highest prices in the recycling market?

C. Resources and market forces
In spite of predictions that metal resources could be depleted within decades, the market prices of most metals, inlcuding lead, have actually fallen since the mid-1970s. Speculate about why the market value of these resources has not reflected their increasing scarcity. For insights, see article by J. Tierney (1990) listed in references below.

D. War and peace
Choose one international political or military conflict from modern or ancient history that was directly related to competition for Earth resources. Describe the nations involved, the resource in dispute, the nature of the conflict, and the ultimate resolution (if any). Cite any outside references you consult.

SUPPLEMENTARY MATERIAL TO EXERCISES

- ## Original "Hubbert" curve (Part II. A. 3., pp. 136-8)
  As noted in the exercises, the Hubbert curve is a 1956 prediction of future petroleum availability. To date, the projected rates of production have proved to be remarkably accurate.

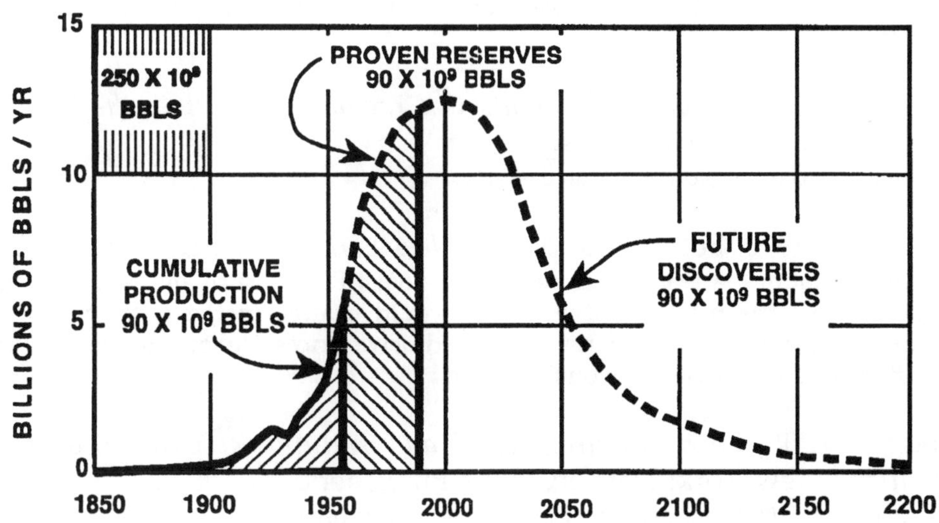

FIGURE S7-1. HUBBERT CURVE FOR WORLD PETROLEUM PRODUCTION
Based on assumed initial reserves of 1250 billion barrels

Source: M. King Hubbert, 1956. Nuclear energy and fossil fuels. *1956 Drilling and Production Practice,* Dallas: American Petroleum Institute. pp. 7-25. Reprinted by permission.

- ## Global petroleum reserves and economic power (Part II. A. 4., p. 138)
  In terms of estimated petroleum reserves, the nations listed rank as follows:
  1. Abu Dhabi: 55 billion barrels
  2. United States: 36 billion barrels
  3. Canada: 9 billion barrels
  4. United Kingdom: 6 billion barrels
  5. Germany: 2 billion barrels
  6, 7: France and Japan : Essentially no reserves.

In recent decades, US petroleum consumption has approximated 30% of world production.

SUGGESTED REFERENCES AND RESOURCES

 Meadows, D.H, Meadows, D.L., Randers, J. and Behrens, W. 1972. *The Limits to Growth*. New York: Universe Books.
The classic work that alerted the public to the issue of resource depletion.

 Kahn, H., W. Brown, and L. Martel, 1976. *The Next 200 Years: A Scenario for America and the World*. New York: William Morrow.
A counterpoint to *Limits to Growth*, presenting a more optimistic view of future resource availability.

 Pestel, E., 1989. *Beyond the Limits to Growth: A Report to the Club of Rome*. New York: Universe Books.
An updated discussion of resource issues.

 Tierney, J., 1990. Betting the planet. *New York Times Magazine*, 2 Dec., p. 52ff.
An account of the 1980 wager between ecologist Paul Ehrlich and economist Julian Simon about how the prices of five metals would change in the coming decade. Ehrlich anticipated that increasing scarcity would drive prices higher. Simon predicted that the prices would decrease and won the bet.

 Tietenberg, T., 1992. *Environmental and Natural Resource Economics, Third Edition*. New York: HarperCollins Publishers, Inc.
An economist's assessment of various models for future resource use.

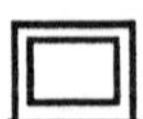 World Resources Institute, 1994. *World Resources: A Guide to the Global Environment*. Oxford: Oxford University Press, 400 pp.
A compendium of information about Earth resources of all types, accompanied by PC-compatible diskettes with databases on consumption and reserves of major commodities. A joint publication of the World Resources Institute and the United Nations Environment and Development programs. Updated biennially.

 Worldwatch Institute publications: *State of the World* yearbooks and Worldwatch papers. Lester Brown, ed.
Rich sources of data on global resource use and abuse, produced by the not-for profit Worldwatch Institute, 1776 Massachusetts Ave. NW, Washington, D.C. 20036-1904 .

 Yergin, D., 1991. The Prize. New York: Simon and Schuster. 877 pp.
A fascinating account of the history of the petroleum industry from its beginnings in the 19th century. This book was the basis for the acclaimed PBS TV series by the same name.

 Youngquist, W., 1990. *Mineral Resources and the Destinies of Nations*. Portland, Oregon: National Book Company.
An examination of the political implications of resource distribution.

 British Petroleum, 1992. *World Total Oil and Gas Reserves* (map). Tulsa, OK: American Association of Petroleum Geologists.
A useful visual summary of estimated oil and gas reserves around the world. Distributed by AAPG Bookstore, P.O. Box 979, Tulsa, OK 74101.

THE IMPERILED EARTH:
**Case study of air and water contamination
at a US Government uranium processing plant**

*Why are radioactive materials dangerous
  to living things?*
*How can the extent of radioactive contamination
  be assessed?*
*Can radioactive waste be stored safely for long
  periods of time?*

In 1951, as part of the Cold War effort to produce processed uranium for defense purposes, the US Government constructed the Feed Materials Production Center at Fernald, Ohio, about 15 km northwest of Cincinnati (Figure 8-1). Uranium was brought to the plant from other Atomic Energy Commission facilities in the form of UF_6 and other compounds. At Fernald, it was converted to uranium metal and made into components for the manufacture of weapons. Production reached its peak in 1960 when 10,000 metric tons of uranium were produced. The plant was closed in 1989. The Fernald site was also the repository for the US stockpile of thorium and for radium wastes produced during the rush to build the first atomic bomb (the Manhattan Project) in the early 1940s. A large quantity of radioactive material is held at Fernald, much of it stored in ways that have allowed contamination of air, soil and water.

Unfortunately, Fernald is only one of many sites in the US where radioactive materials are stored in temporary and inadequate facilities. In this laboratory, you will assess the extent of radioactive contamination at the Fernald plant and consider some of the problems associated with handling and disposal of radioactive waste.

OBJECTIVE

- To gain an understanding of the practical and political difficulties associated with treatment and storage of nuclear waste

OUTLINE

I. Radioactivity revisited
  Modes of radioactive decay
  Decay of uranium isotopes

II. Air and water contamination by radioactive materials at Fernald, Ohio
  Airborne uranium contamination   Airborne radon contamination
  Waterborne uranium contamination   Assessing the costs of cleanup

III. Geologic characteristics of the proposed high-level nuclear waste repository
  at Yucca Mountain, Nevada
  Criteria for selecting a nuclear waste repository
  Geologic controversies at Yucca Mountain

<u>LAB PREVIEW</u> Before coming to lab,
- Review discussion of radioactive decay in Laboratory 2
- Answer as many of the following questions as you can

How much do you know already about radioactivity and radioactive waste?

What is radioactivity? Does it occur naturally?

Give some examples of radioactive elements. Are all isotopes radioactive?

What are the 3 main modes of radioactive decay? Which is potentially most damaging to living things?

How can the direction of groundwater flow be determined? How can the extent of subsurface contamination by a pollutant be assessed?

What is radon? What radioactive decay series produces it? What geologic conditions may lead to high levels of radon in homes?

What characteristics of the Yucca Mountain Nevada site make it an appropriate choice for a nuclear waste repository? What characteristics of the site have caused controversy about its appropriateness?

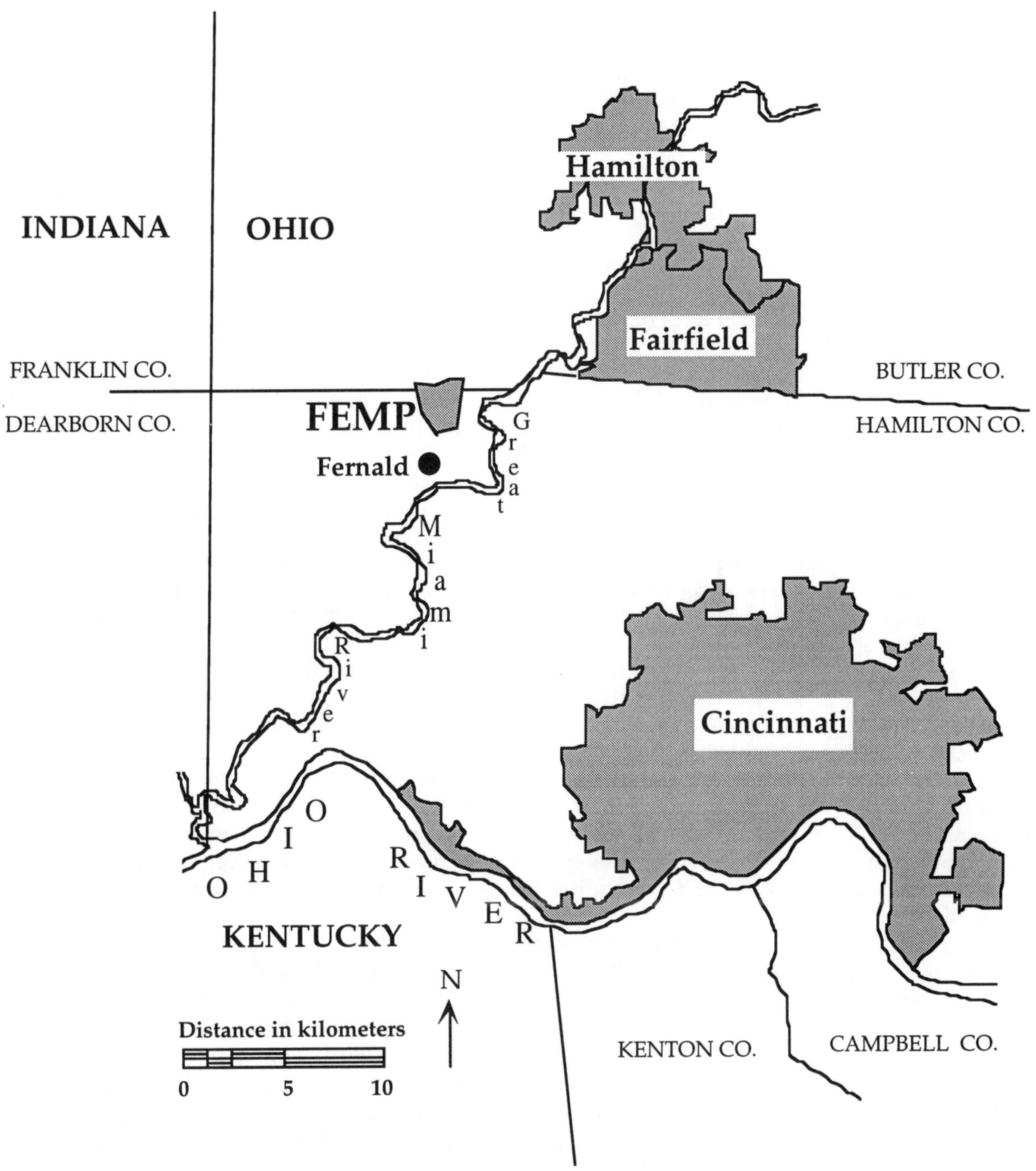

FIGURE 8-1. LOCATION OF THE FEED MATERIALS PRODUCTION CENTER, NOW CALLED THE FERNALD ENVIRONMENTAL MANAGEMENT PROJECT (**FEMP**), SOUTHWESTERN OHIO

I. Radioactivity revisited

In Laboratory 2 (*Earth in Time*) you learned how naturally occurring unstable isotopes can be used to determine absolute ages of rocks. Isotopes of potassium, thorium, uranium and other elements decay at known and constant rates, releasing subatomic particles and heat energy. These radioactive isotopes occur in such low concentrations in most rocks that they pose no real danger to humans. When processed and concentrated, however, radioactive isotopes become hazardous because the subatomic particles they emit can cause <u>mutations</u> in the DNA in living cells.

A. Modes of radioactive decay

There are three main modes of radioactive decay, each of which produces a distinct form of radiation. <u>Alpha decay</u> produces <u>alpha particles</u>, which consist of two protons and two neutrons and thus have a positive charge. Because of their large size and charge, they interact readily with other atoms and do not penetrate solid materials to significant depth. However, high energy alpha particles can do significant damage to living cells because they deposit all their energy in a very small volume of tissue. <u>Beta decay</u> releases electrons, which carry a negative charge. Electrons are much smaller than alpha particles and can travel further in air and into solid material. Because they are smaller and have less charge, however, they cause less concentrated damage to living cells. <u>Gamma decay</u> produces <u>gamma rays</u> -- bundles of electromagnetic energy which, like X-rays, can penetrate some solid materials. Because of their great penetration power, they have the potential to cause significant biological damage.

The intensity of radiation emitted by unstable isotopes in a material is described in terms of <u>activity</u> -- the number of decay events per unit of time. The standard unit of activity is the curie (Ci), named for Marie and Pierre Curie, pioneers in the study of radioactivity. One curie equals a disintegration rate of 37 billion atoms per second. Levels of contamination at Fernald are measured in picocuries (pCi) or 10^{-12} curie. Different radioactive isotopes have different activities. For example, a 1500 kilogram mass of natural uranium (mainly uranium-238 with a half life of 4.5 billion years) has an activity of one curie while just 1 gram of radium-226 (with a half life of 1,620 years) has the same activity.

B. Decay of uranium isotopes

Uranium and radon are the two radioactive elements of greatest environmental concern at Fernald. Uranium is the heaviest of all naturally-occurring elements, with a density of 19.05 g/cm^3. There are two principal isotopes of uranium, uranium-235 (^{235}U) and uranium-238 (^{238}U), with half-lives of 710 million and 4.5 billion years, respectively. Uranium-238 makes up 99.3 % of all uranium in the Earth. The much rarer ^{235}U is the isotope required for generation of nuclear power. The Fernald plant processed mainly ^{238}U.

Uranium-238 is the first isotope in a decay series which ends with the stable lead isotope lead-206 (^{206}Pb) (Table 8-1). One of the unstable daughter isotopes produced in this decay series is radon-222 (^{222}Rn). It is a colorless, odorless <u>noble gas</u> that is heavier than air and has a half-life of only 3.8 days. It decays to a series of short-lived daughters. Serious cell damage can occur if radon or its daughters decay in the lung.

---

mutation:
a change in the genetic code (DNA) within a cell. Some mutations are harmless; others can lead to cancer and birth defects.

alpha decay:
radioactive decay in which an unstable isotope emits a particle with 2 protons and 2 neutrons (**alpha particle**)

beta decay:
radioactive decay in which an unstable isotope emits an electron from its nucleus (**beta particle**)

gamma decay:
radioactive decay in which an unstable isotope emits a form of electromagnetic radiation called a gamma ray

activity:
number of radioactive decay events per unit time

noble gas:
one of the gaseous elements in the last column of the periodic table. With full outer electron shells, they do not bond with other elements

TABLE 8-1.
THE URANIUM-238 DECAY SERIES

STEP #	PARENT	HALF-LIFE	DECAY MODE	DAUGHTER
1	^{238}U	4.47 E+09 yrs	Alpha	^{234}Th
2	^{234}Th	24.1 days	Beta	^{234}Pa
3	^{234}Pa	1.17 minutes	Beta	^{234}U
4	^{234}U	2.44 E+05 yrs	Alpha	^{230}Th
5	^{230}Th	7.7 E+04 yrs	Alpha	^{226}Ra
6	^{226}Ra	1.6 E+03 yrs	Alpha	^{222}Rn
7	^{222}Rn	3.82 days	Alpha	^{218}Po
8	^{218}Po	3.05 minutes	Alpha	^{214}Pb
9	^{214}Pb	26.8 minutes	Beta	^{214}Bi
10	^{214}Bi	19.8 minutes	Beta	^{214}Po
11	^{214}Po	1.64 E-04 secs.	Alpha	^{210}Pb
12	^{210}Pb	22.3 years	Beta	^{210}Bi
13	^{210}Bi	5.01 days	Beta	^{210}Po
14	^{210}Po	38.4 days	Alpha	^{206}Pb (stable)

1. What would you expect to be the relationship between half-life and activity for unstable isotopes? That is, which isotopes would have higher activities -- those with long half-lives or those with short half-lives? Explain.

atomic mass number:
sum of the number of protons and neutrons in the nucleus of an atom

2. The symbol or abbreviation for an isotope (e.g. ^{235}U) includes the symbol for the element (U) and a superscript indicating the <u>atomic mass number</u> of the isotope -- the sum of the number of protons and neutrons in the nucleus (Lab 2). Why does the atomic mass number of the daughter products decrease progressively through the ^{238}U decay series (Table 8-1)?

3. For one step in the ^{238}U decay series, show that the decrease in atomic mass number is consistent with the decay mode indicated.

II. Air and water contamination by radioactive materials at Fernald, OH

Careless handling and storage of the uranium processed at Fernald lead to dispersal of radioactive materials by air and water.

A. Airborne uranium contamination

From 1953 to 1975, uranium oxide from the milling process at Fernald escaped into the air through holes in poorly designed filter bags. Available data suggest that a minimum of 136,000 kilograms (300,000 lbs.) of uranium oxide were released into the air during the plant's years of production.

Another source of airborne contamination are uranium-bearing wastes stored outside the plant in large pits. The dried residue at the surface of these pits can be picked up and transported by the wind. The quantity of uranium which has escaped in this way is unknown.

It is troubling to consider how widely this uranium may have been dispersed. The largest particles are thought to have fallen within the plant site boundaries because of the high density of uranium. Citizens living near the plant, however, have been understandably concerned about the extent of contamination. In order to assess the degree of contamination from airborne uranium on and around the facility, surface soil samples were collected in 1989 and analyzed for their activity. Figure 8-2 illustrates the results of this study. Uncontaminated dry soil in southwestern Ohio has natural activity levels of 2.2 to 3.0 pCi/gram.

contour line:
line joining points on a map where some quantity (e.g. elevation, barometric pressure) has the same value

1. Show how uranium activities in the soil vary in the Fernald area by drawing <u>contour lines</u> on the map in Figure 8-2. Draw lines separating areas that have activities of a) less than 3, b) 3-10 and c) more than 10 pCi/g. (Consider 3 pCi/g the natural background level of radioactivity). In effect, you are making a "topographic" map of uranium activities.

Is there any evidence for uranium contamination of these soils? If so, where? Explain.

2. Table 8-2 contains information about wind directions in the Fernald area in 1990. These data can be more easily interpreted by creating a "rose diagram" -- a kind of circular bar graph -- in which the length of each "petal" represents the percent of time the wind blew from a particular direction. Complete the rose diagram below Table 8-2. The westerly petal has already been drawn in. (Alternatively, the diagram could be created with a spreadsheet program if the option of a "polar" plot is available. Note that in some programs, directional information must be entered in radians rather than in degrees).

Do you think that the observed pattern of soil contamination was due to dispersal by wind? Explain.

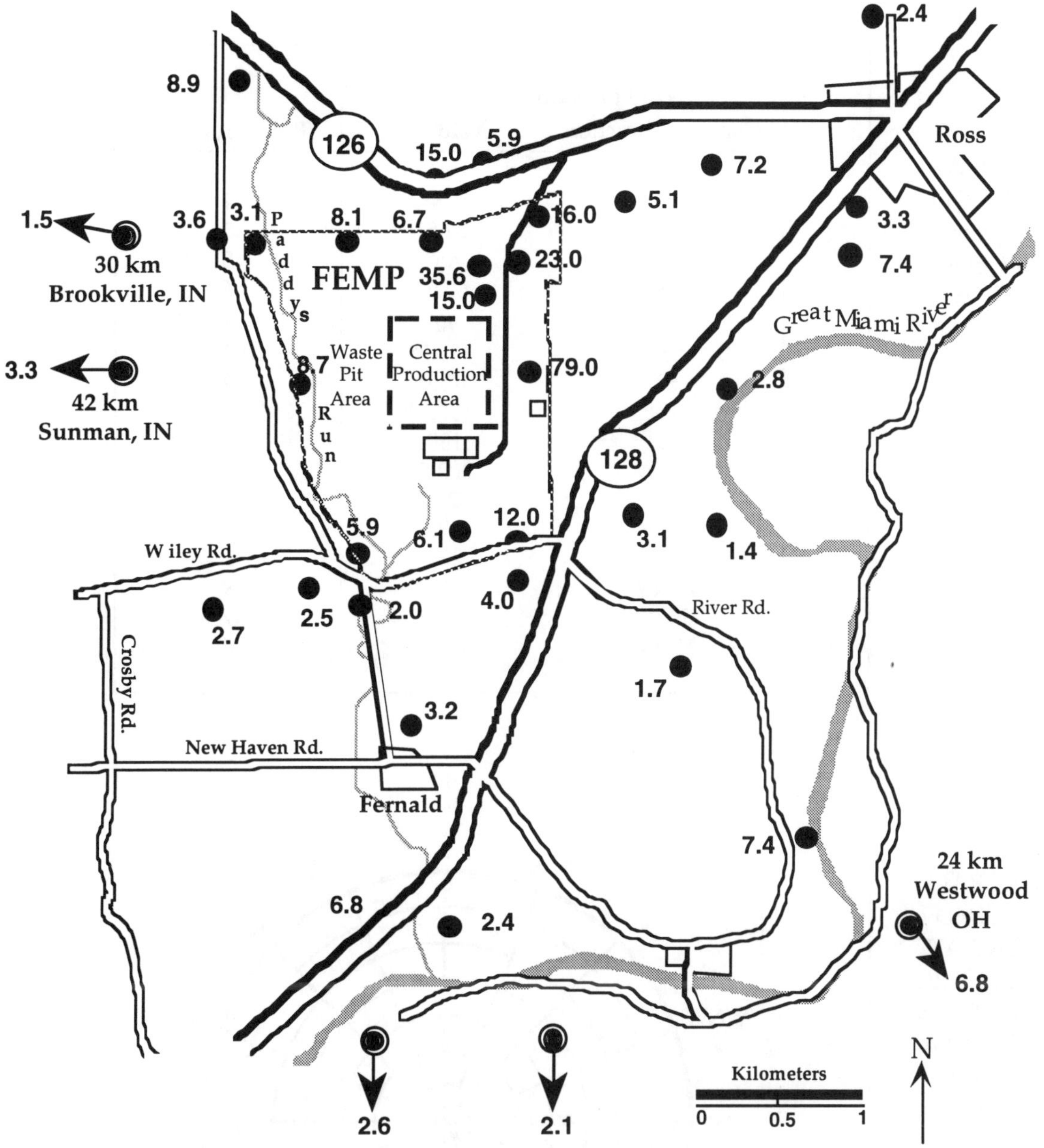

EXPLANATION

● Soil sampling location	------- Perimeter of plant grounds
◄—◉ Distance from production area to sampling locations off map	— — — Perimeter of production area

FIGURE 8-2. URANIUM ACTIVITIES IN SOIL (pCi/g) NEAR THE FERNALD PLANT IN 1989
Source: US Department of Energy

TABLE 8-2.
1990 WIND DIRECTIONS AT FERNALD, OHIO (directions *from which* wind blew)
Source: US DOE

| Wind direction at 60 m: | | |
Compass direction	Azimuth	Percent of time
N	0°/360°	4
NNE	22.5°	5
NE	45°	5
ENE	67.5°	6
E	90°	4
ESE	112.5°	2
SE	135°	2
SSE	157.5°	3
S	180°	4
SSW	202.5°	9
SW	225	13
WSW	247.5°	12
W	270°	11
WNW	292.5°	9
NW	315°	7
NNW	337.5°	4

FIGURE 8-3.
ROSE DIAGRAM OF 1990 WIND DIRECTIONS AT FERNALD, OHIO (to be completed by student). Concentric circles represent 2% time intervals.

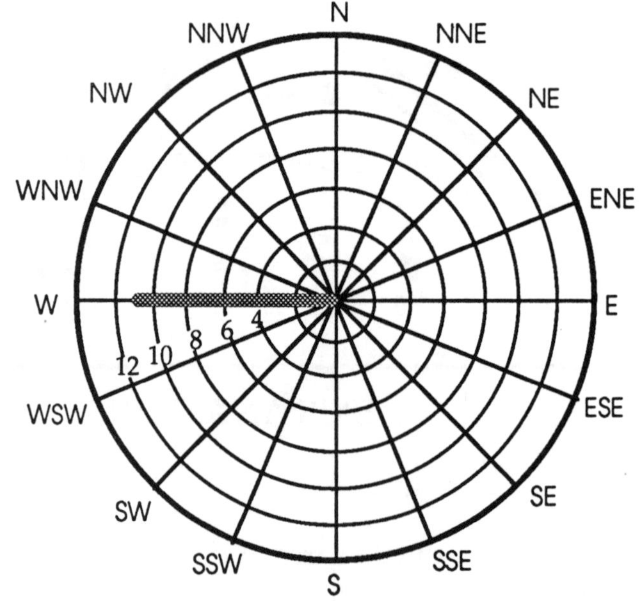

terrace:
flat land surface adjacent to a river, representing an abandoned floodplain formed at a time when the river flowed at a higher elevation

aquifer:
a permeable rock or sediment layer through which groundwater flows

plume:
area of subsurface groundwater contamination that originates from an identifiable source

Groundwater extraction wells have been installed at the leading edge of the plume. The wells must pump continuously to prevent contamination of the regional aquifer.

FIGURE 8-4.
EAST-WEST CROSS SECTION THROUGH THE GREAT MIAMI RIVER VALLEY AT FERNALD, OHIO
Line of section shown in Fig. 8-5. Vertical scale exaggerated by a factor of 20.

B. Waterborne uranium contamination

Water, both on and under the surface, has also carried uranium from the Fernald site. To understand the dispersal of uranium by water, we need to consider the geology of the area.

The Fernald plant is located on a river <u>terrace</u> 15.2 meters above the present floodplain of the Great Miami River. As shown in the cross section (Fig. 8-4), nearly 15 meters of clay-rich material (glacial till and lake deposits) overlie about 46 meters of sand and gravel (glacial meltwater outwash deposits). This sand and gravel layer is one of the most productive <u>aquifers</u> (Lab 4) in Ohio. Note that Paddy's Run, a creek that flows through the plant site, has cut into a sandy lens that is connected to the aquifer beneath.

The potential for groundwater contamination is considerable given the large amounts of uranium on the surface. Four unlined pits were filled with liquid wastes, including uranium compounds, during the plant's 40 years of production. The pits range in size from about 2500 to 50,000 m^2 and in depth from 4 to 10 m. Surprisingly, serious pollution of groundwater in the sand-gravel aquifer beneath the Fernald site has been documented at only two places to date. One lies directly beneath the waste-disposal pits and the second near the SW corner of the plant grounds (Fig. 8-5).

Uranium contamination of groundwater has occurred, however, beyond the boundaries of government property, near the southwest corner of the plant. This area of polluted ground water has been designated the "south <u>plume</u>". Background concentrations of uranium in groundwater in the area range from about 0.68 pCi/l to about 2.0 pCi/l, but concentrations as high as 190 pCi/l have been recorded in the plume (Fig. 8-5). Although no official maximum concentrations have been set by the US Environmental Protection Agency (EPA), 13.5 pCi/l has been proposed as the drinking water standard.

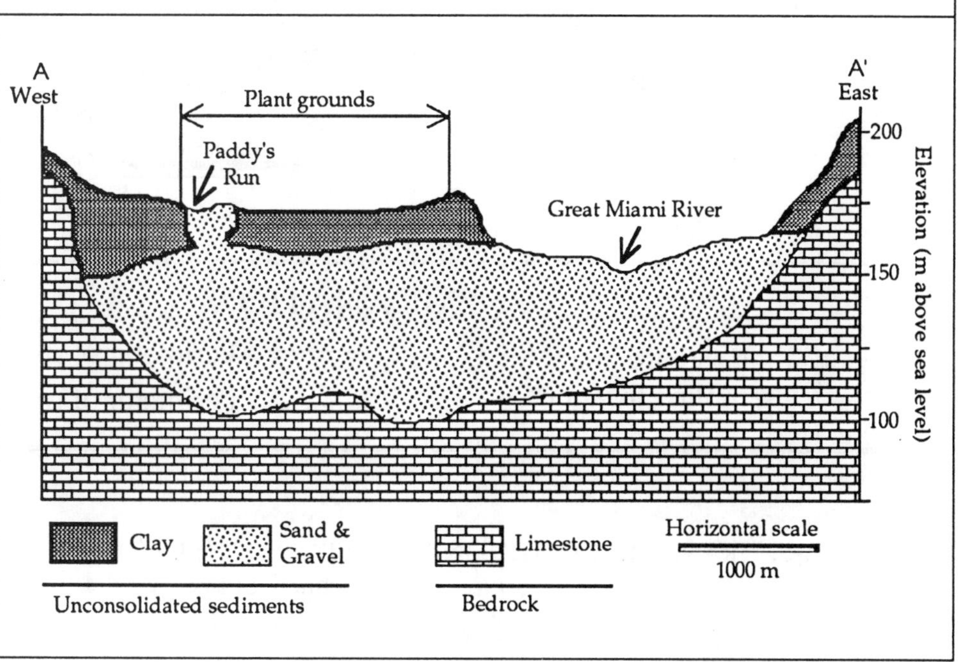

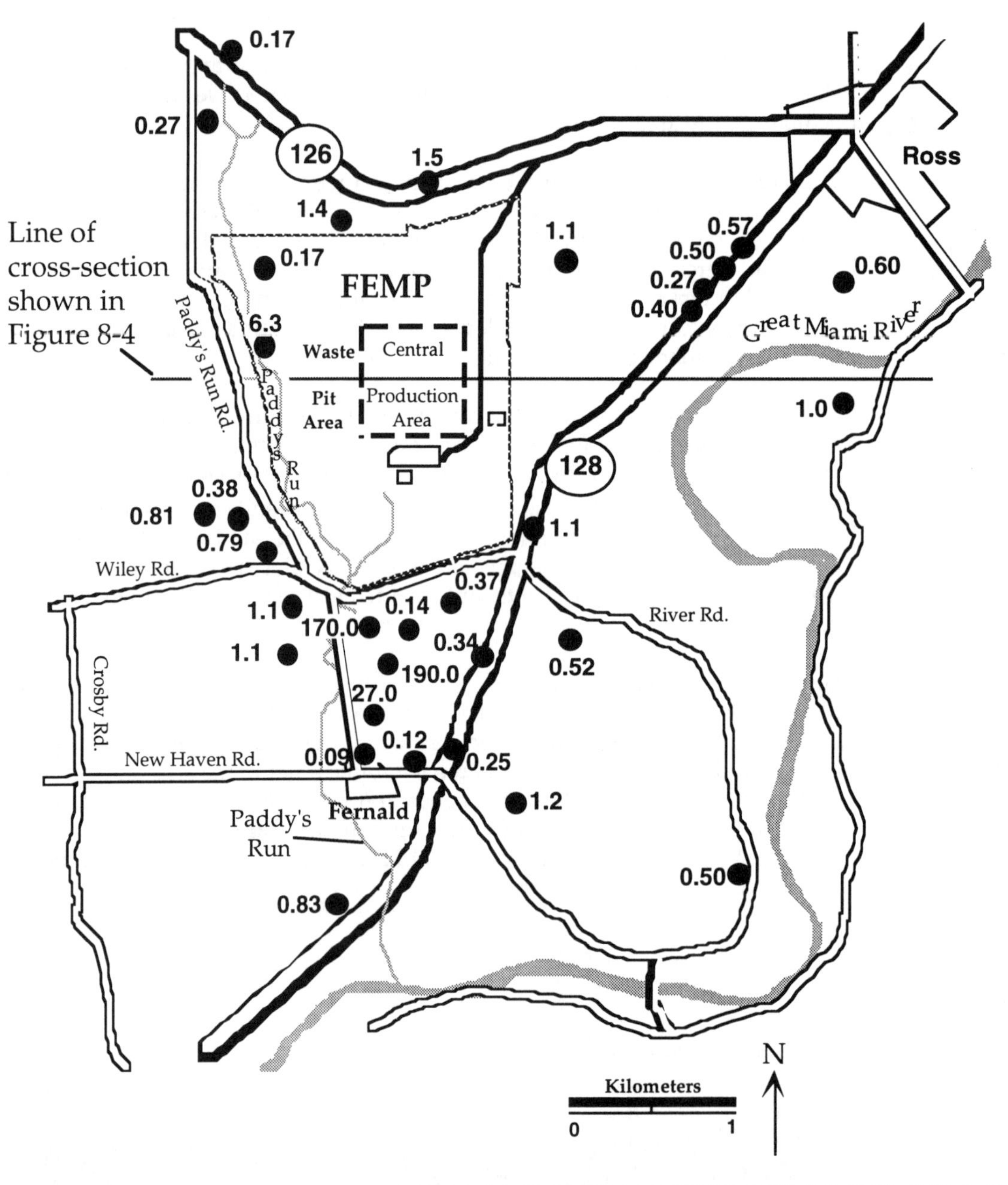

FIGURE 8-5. URANIUM ACTIVITIES IN GROUNDWATER (pCi/l) IN THE FERNALD AREA, 1989
Source: US Department of Energy

point source:
location that can
be considered the
origin of a
spreading pollutant

water table:
top of the zone in
the subsurface
where all pore
spaces in sediments
and rocks are
filled with water
(the zone of
saturation)

**equipotential
lines:**
lines connecting
points on a map
where the water
table is at the
same elevation

FIGURE 8-6.
EQUIPOTENTIAL
LINES AND FLOW
DIRECTIONS FOR
A HYPOTHETICAL
AQUIFER

1. Why do you think surface contaminants have not yet reached the sand and gravel aquifer at most places beneath the plant site (Figs. 8-4, 8-5)?

What might explain the contamination in the Wiley Road area?

2. To protect the water supplies of nearby residents, it is important to establish the direction in which uranium-contaminated groundwater is flowing. There are two approaches to this.

a) One method is to assume that the contamination originates from a <u>point source</u> and to study the distribution of already contaminated wells. Based on the pattern of contamination shown in Figure 8-5, in what direction does groundwater flow in the area between the plant site and New Haven Road? Explain.

b) Predicting the direction of groundwater flow *before* contamination occurs, requires knowledge of the "topography" of the regional <u>water table</u>. This is done by recording the elevation of water in wells that penetrate the aquifer. If enough data are available, lines of equal water table elevation, <u>equipotential lines</u>, can be drawn, creating a subsurface map of the water table.

Like surface water, groundwater flows in response to gravity, down the 'slope' of the water table, perpendicular to equipotential lines (Fig. 8-6). Figure 8-7 shows the location of water wells on and off the Fernald site and the elevation of water (in feet above sea level) in each well on 1 April 1990. Draw equipotential lines at two foot intervals on the map, then add arrows showing the direction of groundwater flow. What is the principal direction of groundwater flow on and near the plant site?

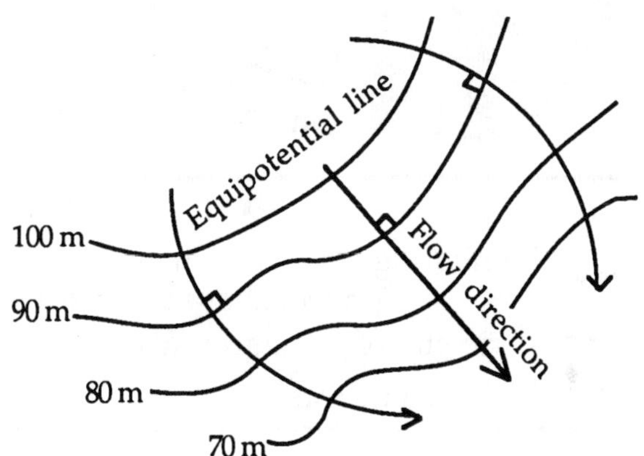

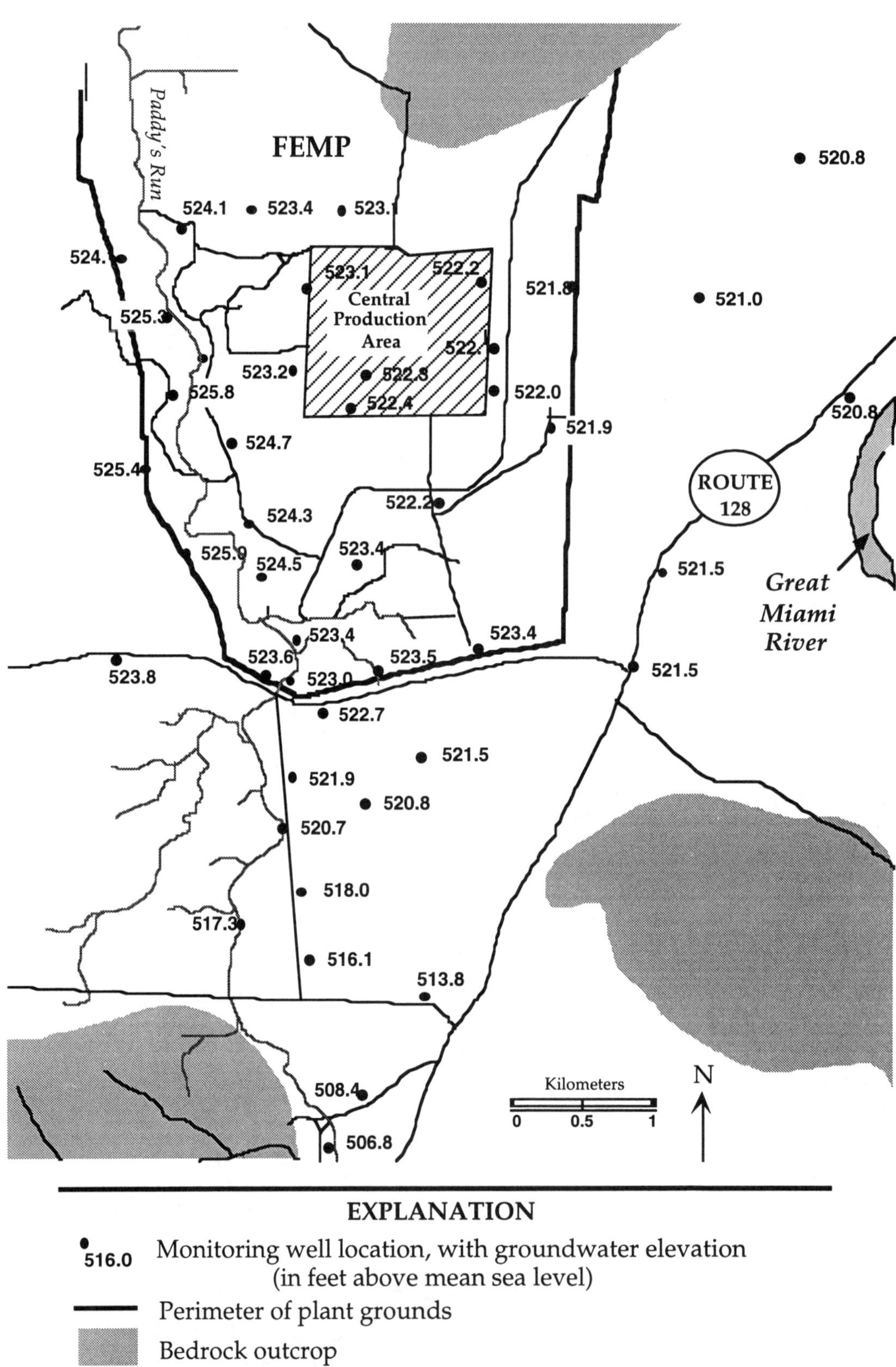

EXPLANATION

FIGURE 8-7. WATER TABLE ELEVATIONS IN THE FERNALD AREA   *Source:* US DOE

Natural radon levels are highest in areas where the bedrock is composed of granite, which typically has higher concentrations of uranium than other rock types.

Houses with basements are more likely to have higher radon levels than those without.

C. Airborne radon contamination

The health hazards associated with breathing radon gas (a daughter isotope in the ^{238}U decay series; Table 8-1) have received considerable media attention in recent years. In homes built on bedrock containing relatively high concentrations of uranium, even natural levels of radon can exceed recommended limits.

At Fernald, radon is being produced in large quantities by the decay of waste radium-226 (^{226}Ra) stored in two deteriorating silos. Until recently, radon gas was escaping through cracks in the cement roofs of the structures. Radon activity in the vicinity of the silos is typically about 0.74 pCi/l, well below the 4.0 pCi/l limit set by the US EPA for airborne radon in the home. At times, however, radon activities have risen as high as 740 pCi/l in the vicinity of the silos.

Figure 8-8 shows the record of a radon detector at ground level 50 m SW of the silos at 1 a.m. on 12 - 18 February, 1991. Figure 8-9 is a plot of the difference in temperatures (ΔT) at heights of 60 m and 10 m above the ground during the same time period. (A positive ΔT value means that it was warmer at 60 m than at 10 m). Figure 8-10 shows wind speeds at the site during this period.

What weather conditions were associated with peak radon concentrations at ground level?

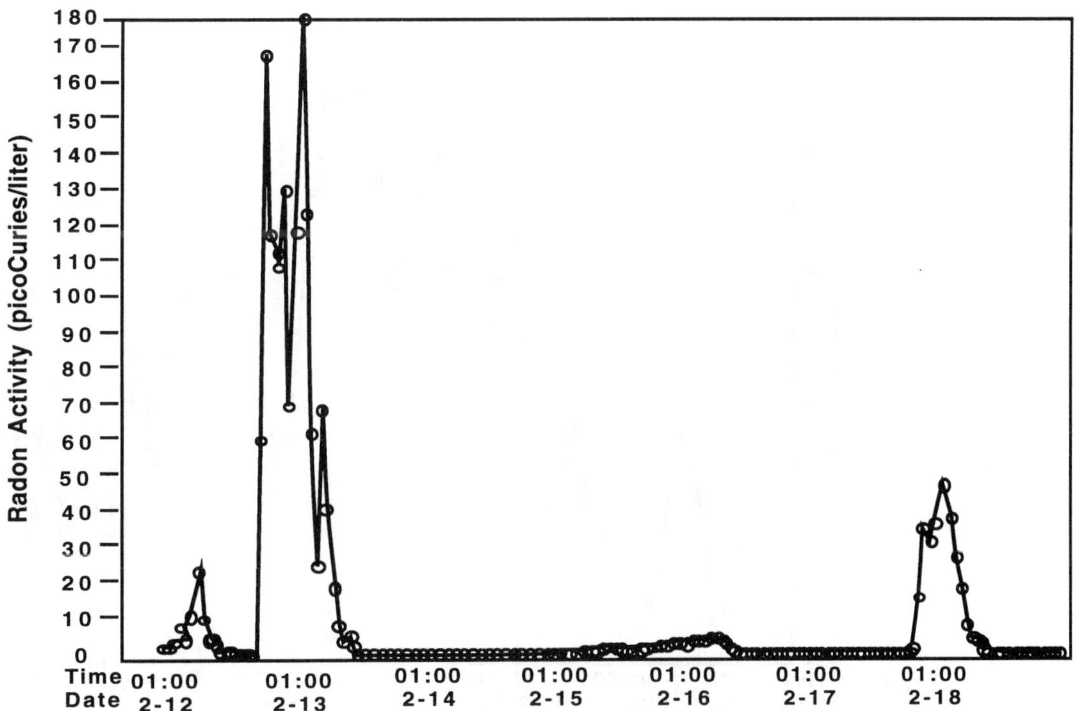

FIGURE 8-8. RADON ACTIVITIES NEAR THE FERNALD SILOS, 12-18 FEBRUARY, 1991

FIGURE 8-9.
DIFFERENCE IN
TEMPERATURES AT
60 M AND 10 M ,
FERNALD, OHIO,
12-18 FEBRUARY
1991

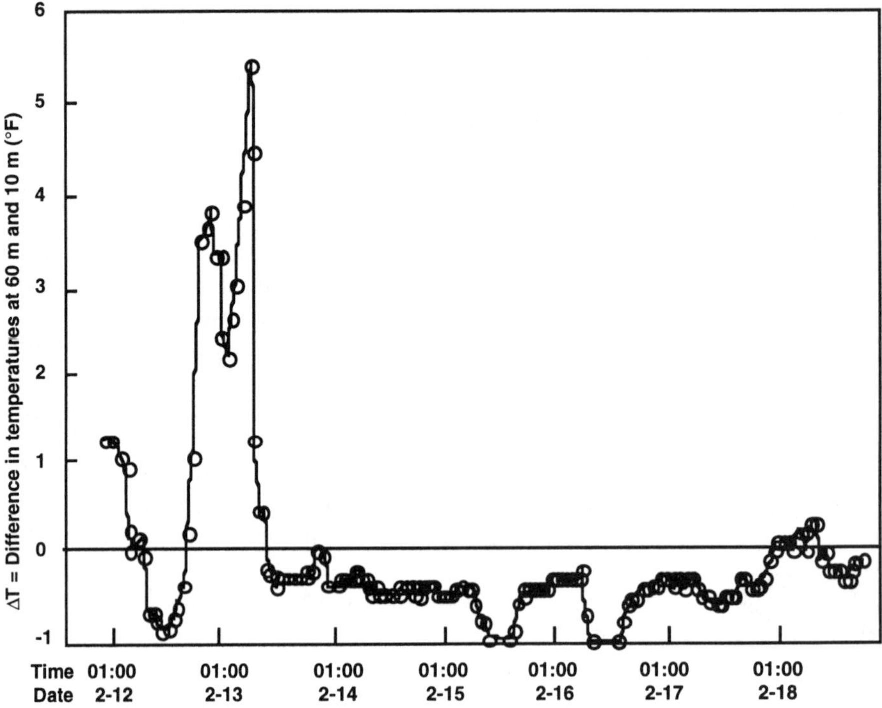

FIGURE 8-10.
WIND SPEEDS,
FERNALD, OHIO,
12-18 FEBRUARY
1991

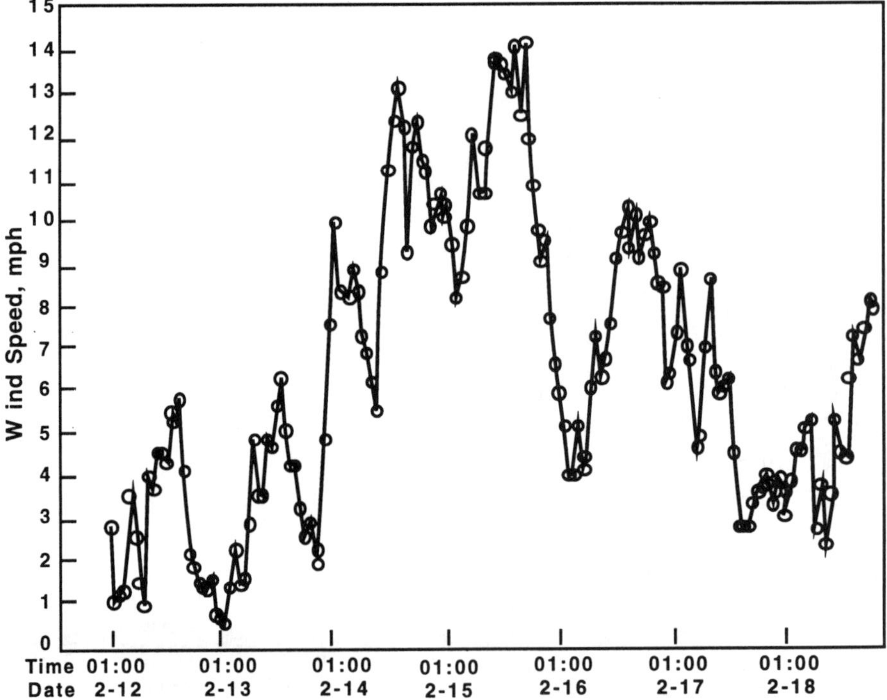

D. Assessing the costs of cleanup

Serious contamination of the Fernald site occurred during the 40 years the plant was active, and wastes stored at the site continue to threaten air, water and soil quality in the surrounding area. The preceding exercises have considered only the most serious environmental problems at Fernald. Others include leakage from the waste-storage pits, escape of radioactive thorium, and pollution of groundwater by organic solvents. The estimated cost of cleaning up the site is more than $5 billion, and it is unlikely that it will ever be restored to an uncontaminated condition. It has been suggested that the site simply be abandoned after enough work is performed to ensure that pollutants do not migrate offsite.

1. What do you think should done at sites like Fernald? Should comprehensive cleanup efforts be undertaken? Why or why not? Should area residents be compensated for the potential threats to their property and health? If so, how? On a separate sheet, write a short statement of your opinions. Justify your stance.

remediation: cleanup and restoration of a polluted site

2. Although the quantity of radioactive waste at Fernald makes it a particularly problematic site, it is only one of dozens across the US where defense-related toxic and radioactive materials pose environmental threats. Table 8-3 lists military sites where radioactive wastes await permanent disposal. Is there a site in your region? Has remediation begun there? What are the projected costs of cleanup?

TABLE 8-3.
U.S. NUCLEAR WEAPONS PRODUCTION FACILITIES WITH INADEQUATELY STORED RADIOACTIVE MATERIALS
Sites where estimated cleanup costs exceed $1 billion are marked "$". Nuclear test sites are marked "#".

Source: Physicians for Social Responsibility and Military Toxics Project, 1993. *Covering the Map: Military Pollution Sites in the US.*

California:	**Missouri:**	**New York:**
Livermore Natl Laboratory	Kansas City Plant	Ashland sites
Colorado:	St. Louis Downtown site	Linde Center
# Project Rio Blanco site	St. Louis Airport site	Knolls Atomic Power Lab
# Project Rulison site	**New Jersey:**	Niagara Falls Storage site
$ Rocky Flats plant	Kellex/Pierpont	Seaway Industrial Park
Florida:	Middlesex Landfill	**Ohio:**
Pinellas plant	Middlesex Sampling Plant	$ Fernald
Idaho:	New Brunswick site	Luckey site
$ Idaho Natl Engineering Laboratory	**Nevada:**	Mound plant
	# Nevada Test Site	Oxford site
Illinois:	**Oregon:**	Painesville site
Granite City Steel	Albany Research Center	Portsmouth Diffusion plant
Madison site	**Pennsylvania:**	**South Carolina:**
Kentucky:	Aliquippa Forge	$ Savannah River site
Paducah Diffusion plant	Springdale site	**Tennessee**
Maryland:	**New Mexico:**	Elza Gate site
W.R. Grace & Company	$ Los Alamos Natl Lab	$ Oak Ridge Natl Lab
Massachusetts:	# Project Gasbuggy Site	$ Oak Ridge K-25 site
Chapman Valve	# Project Gnome Site	Oak Ridge Y-12 plant
Shpack Landfill	Sandia National Lab	**Texas:**
Ventron Corporation	South Valley site	Pantex plant
Michigan	# Trinity site	**Washington**
General Motors site	$ Waste Isolation Pilot Plant (WIPP)	$ Hanford Nuclear Res.
Mississippi:		
Tatum Dome		

high level radioactive waste: highly radioactive materials from nuclear power plants and weapons development

Although the contamination problems at Fernald are serious, most of the radioactive materials at the site are classified as low-level waste.

Do you think waste should be stored above or below the level of the water table?

What characteristics of nuclear waste make it especially difficult to store safely for long periods?

FIGURE 8-11.
LOCATION AND GEOLOGIC SETTING OF THE PROPOSED NUCLEAR WASTE REPOSITORY AT YUCCA MTN., NEVADA
Source: US DOE, Yucca Mtn Project

III. Geologic characteristics of the proposed high-level nuclear waste repository at Yucca Mountain, Nevada

Although many radioactive waste sites are legacies of Cold War weapons production, much of the <u>high-level</u> radioactive waste in the U.S. is in the form of spent fuel rods from commercial nuclear power plants. More than 20,000 metric tons of this highly radioactive material has accumulated in the 40 years of nuclear power generation, but there has never been any provision for its long-term storage. At most of the country's 109 nuclear plants, the spent fuel rods are being kept "temporarily" in cooling pools at the reactor sites. Recognizing that the radioactive waste problem could only grow more and more intractable with time, the US Congress directed the Department of Energy (DOE) in 1982 to begin a search for a national high-level waste repository site.

After several years of study and contentious political debate, DOE proposed three possible sites: Hanford, Washington; Deaf Smith County, Texas; and Yucca Mountain, Nevada (Fig. 8-11). In 1986, Congress ordered DOE to focus its site characterization efforts on Yucca Mountain alone. Geological analysis of the site is to last through the year 2001, at an estimated total cost of $6 billion.

A. Criteria for selecting a nuclear waste repository

1. What political and social considerations must be taken into account in identifying sites for waste disposal (whether for nuclear waste or ordinary garbage)?

2. What environmental and geological considerations must be taken into account in siting a nuclear waste repository? Federal legislation mandates that high-level waste must be isolated from the environment for 10,000 years.

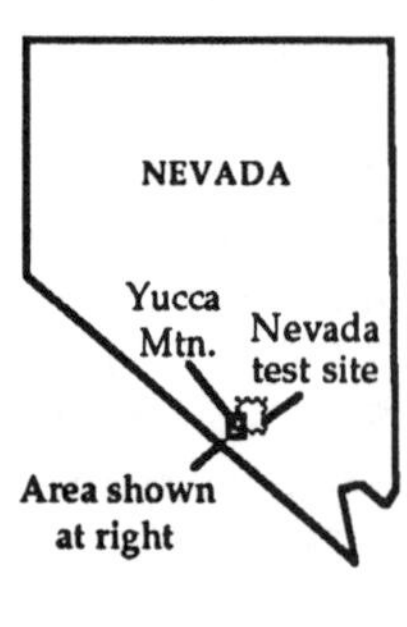

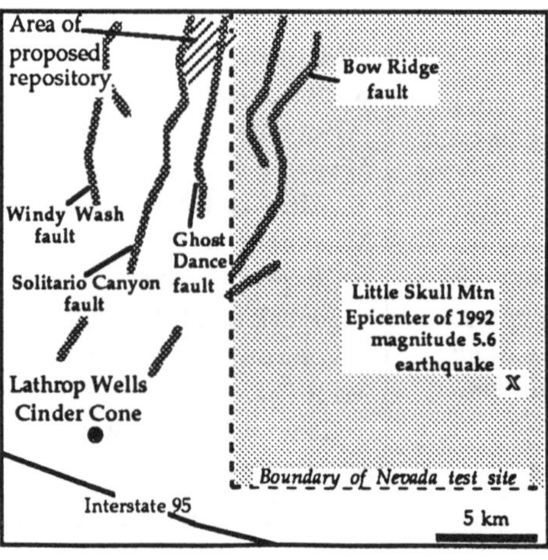

tuff:
rock formed from compacted volcanic ash. If ash is still hot when it accumulates, tuff may be fused or **'welded'**

permeability:
measure of the ease with which fluids can move through sediment or rock (Lab 4)

B. Geologic controversies at Yucca Mountain

Several characteristics of the Yucca Mountain site make it an attractive candidate for a nuclear waste repository. First, it lies in a virtually unpopulated area, which would minimize potential risks to human health (and political resistance to the project). Second, the bedrock is <u>welded tuff,</u> a highly <u>impermeable</u> rock composed of fused volcanic ash flows (review discussion of volcanic rocks in Lab 3). Because the rock was formed at high temperatures, it would be able to withstand the heat generated over time by the waste itself, and its low permeability would reduce the chances of waste leaking out of the storage site. Third, the area receives only a few centimeters of rain each year (it is only 70 km from Death Valley), and the water table in the region is almost 700 m deep (Fig. 8-12). These conditions would decrease the likelihood that groundwater could carry contaminants away from the site.

Other geologic characteristics of the site, however, have raised concerns about its suitability. These are the main focus of the ongoing studies at Yucca Mountain.

FIGURE 8-12.
GEOLOGIC
CROSS SECTION
THROUGH
YUCCA MTN.
Source: US DOE,
Yucca Mtn Project

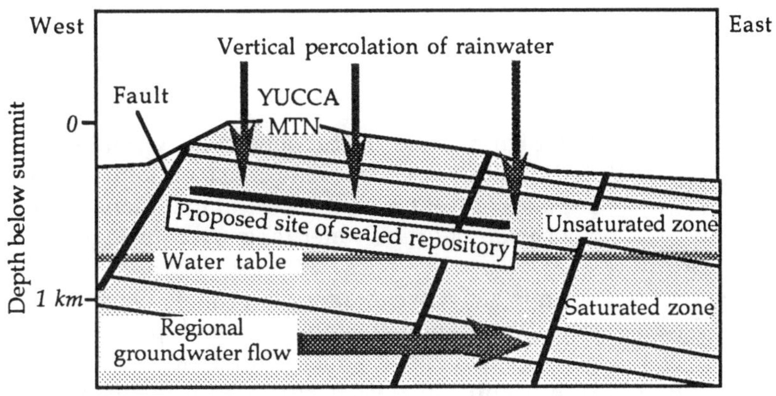

cinder cone:
small, conical volcano formed from pebble-sized pieces of quickly cooled lava. Most cinder cones are 100m high or less

Yucca Mountain lies in a geologic province called the Basin and Range, an area of regional lithospheric extension.

1. Recent volcanic activity

The volcanic tuff that forms Yucca Mountain itself is 12 to 15 million years old, and there is no concern about reactivation of the explosive volcanoes from which the ash flows emanated. There are, however, several small volcanic <u>cinder cones</u> that have been active as recently as 10,000 years ago. One of these, Lathrop Wells volcano (Fig. 8-11), is only 18 km from the proposed repository site. DOE geologists maintain that these are isolated volcanic centers and that the probability of an eruption directly beneath the repository is very low.

a) What does the highly symmetrical shape of cinder cones suggest about the manner in which they grow?

b) Review Figures 3-2 and 3-3, Laboratory 3. Does Yucca Mountain lie on a plate boundary? If so, what type? If not, where are the nearest plate boundaries and what types are they? Is the area part of a major volcanic belt?

2. Continuing seismic activity

Moderate-sized earthquakes are common in the Yucca Mountain region. In June 1992, a magnitude 5.6 quake, the largest in a century, occurred at Little Skull Mountain (Fig. 8-12), only 19 km from the proposed repository. Several buildings at the DOE study facility were damaged, and new concerns were raised about the possibility that a major earthquake could seriously compromise the repository.

a) What geologic characteristics of the Yucca Mountain site indicate that it has had a long geologic history of seismic activity (Figs. 8-11 and 8-12)?

b) What approaches could be used to reconstruct the frequency of major earthquakes in the region in the recent geologic past (last 10,000 years)?

3. Past and future fluctuations in the water table

Because the nuclear waste must be kept isolated from regional groundwater supplies for centuries, it is critical to understand what factors can cause changes in the elevation of the water table over geologic time.

Seismic activity can alter water table elevations because rock fracture and fault slip can change the porosity and permeability of subsurface rocks. Some geologists believe that groundwater levels at Yucca Mountain could rise significantly in the aftermath of a major earthquake. They site calcite <u>veins</u> that run through surface outcrops of the volcanic tuff as possible evidence for much higher groundwater levels. Groundwater typically carries minerals in solution and the dissolved calcite may have been precipitated in cracks and pores in the rock as the water evaporated. Alternatively, the veins could have been formed by surface water percolating downward. In this case, they would shed no light on past levels of the water table.

a) Study the specimens of rock with calcite veins, then describe and sketch them.

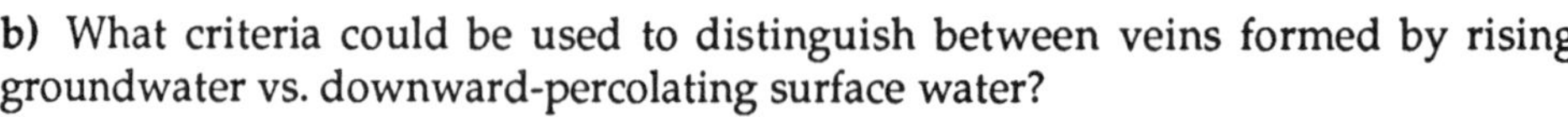

b) What criteria could be used to distinguish between veins formed by rising groundwater vs. downward-percolating surface water?

c) Besides seismic activity, what other factors could cause the elevation of the water table to change on a time scale of decades or centuries? Explain.

IV. Synthesis

A. Preview review
Return to the Lab Preview and complete any questions you were unable to answer before.

B. When we've been here ten thousand years...
Current US legislation requires that the nation's high-level nuclear waste repository must be designed to last 10,000 years. How long is 10,000 years? To put this in perspective, answer the following questions. What major period in Earth history was just ending 10,000 years ago? At 30 years each, how many human generations does 10,000 years represent? How many years have the most enduring human civilizations lasted? How long does it typically take for a language to evolve to the point where speakers from two different time periods cannot understand each other? Neptunium 237 (^{237}Np), one of the unstable isotopes in high-level radioactive waste has a half-life of 2.14 million years. How many 10,000-year periods must pass before only half of the original radioactive parent material remains? Explain your calculation.

C. Taking it all into account
Based on what you have learned in this lab and previous ones, write a statement comparing the "hidden costs" of nuclear and fossil fuel-derived energy. Who should pay the costs of waste disposal and environmental degradation? What regulatory and taxation strategies would be most effective in encouraging responsible energy use?

SUGGESTED REFERENCES AND RESOURCES

Grossman, D. & Shulman, S., 1994. Verdict at Yucca Mtn. *Earth*, v.3 n.2, p. 54-63.
A readable and objective account of ten years of geologic studies at Yucca Mountain.

Johnson, C. & Hummel, P., 1991. Yucca Mtn., NV. *Geotimes*, v.37 n. 8, p. 14-16.
Mattson, S., Younker, J., Bjerstedt, T., and Bergquist, J., 1992. Assessing Yucca Mountain's Natural Resources. *Geotimes*, v. 38 n. 1, p. 18-20.
A pair of articles presenting opposing views on the possibility that natural mineral resources at Yucca Mountain could become attractive targets for future exploitation and that their extraction could lead to leakage of radioactive materials from the site.

Kimball, D., Siegel, L., and Tyler, P., 1993. Covering the Map: A Survey of Military Pollution Sites in the United States. Physicians for Social Responsibility and Military Toxics Project. Unpaginated.
A sobering summary of defense-related toxic and radioactive waste sites in the US compiled by Physicians for Social Responsibility, 1000 16th St. NW, Suite 810, Washington, DC 20036 and the Military Toxics Project, Rte. 1, Box 2020, Litchfield, ME 04350. Phone: 207/268-4071.

Lenssen, N., 1991. Nuclear Waste: The problem that won't go away. *Worldwatch Paper*, n. 106. Washington DC: Worldwatch Institute. 64 pp.
An assessment of the nuclear waste problem at the global scale.

National Research Council, 1992. Groundwater at Yucca Mountain: How High Can It Rise? Washington, D.C.: National Academy Press. 231 pp.
A thorough examination of the proposal that seismic activity at Yucca Mountain could cause significant upwelling of groundwater. The report concludes that large changes in water table levels are unlikely but offers recommendations for further analysis of the issue.

APPENDIX I: SPREADSHEET INSTRUCTIONS

Exercises marked by the symbol ▮▮▮ in the text may be completed using spreadsheet programs. Detailed instructions are given below for four commonly used programs: Excel (PCs and Macintosh); Lotus 1-2-3 (for PCs); Quattro Pro (PCs) and Wingz (PCs and Macintosh). These instructions are written for students who have no previous experience with spreadsheets. Progressive mastery of skills is assumed, however, and directions are provided only when new procedures are required.

SPREADSHEET BASICS (all programs)

Spreadsheets are programs that facilitate quantitative analysis and graphical representation of data. Originally developed for accounting purposes, they are now widely used in many fields. The spreadsheet skills you acquire in this course will be useful elsewhere.

Starting the program and opening files: Using the mouse, position the cursor over the program icon and click the mouse button twice rapidly (use the left mouse button for PCs). The spreadsheet program will open and a blank page or "worksheet" will appear. This is a new file into which data can be entered (see below). Additional blank worksheets can be opened by using the mouse to select *New* from the **FILE** menu. To open an existing file, choose *Open* from the **FILE** menu, then indicate the drive and directory (folder) in which the file occurs and click 'OK'. (For PCs, 'A' and 'B' are generally floppy disk drives; 'C' is the computer's hard drive).

Worksheet layout: Spreadsheet worksheets consist of cells arranged in rows and columns. Cells are identified by their row and column labels (for example, B2 = Column B, Row 2). Practice moving from one cell to another using: 1) the mouse; 2) the Return/Enter key; 3) the tab key; and 4) the arrow keys. You can adjust column widths by placing the cursor on the divider line at the top of a column, holding down the mouse button, and dragging the line left or right.

Entering and editing data: Select a cell by clicking on it. Type something (numbers or letters) in this cell. The contents of the cell are displayed in the "entry bar" at the top of the screen. To change the contents of a cell, use the mouse to move the cursor to the correct position in the entry bar, click, then type. To delete the contents of a cell entirely, select the cell by clicking on it, then press the Delete key.

Copying, cutting and pasting entries: Select one or more cells by clicking and dragging the mouse until all that you want to copy or cut are highlighted. If you want to move the contents of a cell from one place to another, select *Cut* from the **EDIT** menu, then click on the new cell location where you want the entries to begin and select *Paste* from the **EDIT** menu. To copy cell contents to new cells without deleting them from their original location, select *Copy* instead of *Cut* from the **EDIT** menu.

Saving a worksheet: From the **FILE** menu select *Save As*. Indicate the drive (and, if applicable, directory or folder) to which the file should be saved and the name the file is to be given. For example, for PCs 'A:\OPUS' (no quotes) saves the file under the name 'OPUS' to a disk in the computer's 'A' drive.

LABORATORY 1: *THE EARTH IN SPACE* 

LAB 1/ Exercise I.B.2., p. 4: Planetary spacing

- **Microsoft Excel (v. 4.0 for Windows; v. 4.0 for Macintosh):**

  > Open Excel

  > To enter data:

    - In cells A1 and B1 respectively, type the words 'Planets' and 'Distance' (no quotes). These will act as titles for your columns of data.

    - In cells A2 to A10, enter the names of the planets, in order of increasing distance from the Sun.

    - In cells B2 to B10, enter the planets' distances from the Sun (Table 1-2) as logarithms. To do this, type '=LOG(distance)' in each cell (no quotes). Numbers in scientific notation must be entered in modified form. For example, 5.79×10^7 would be 5.79E+07. So the full entry in B2 should be: '=LOG(5.79E+07)'.

  > To create a graph:

    - First, highlight the range of cells to be graphed (cells A2 through B10).

    - Next, click on the **Chart Wizard** (Excel v. 4.0) or the **Graph** button (earlier versions) on the upper right side of the icon bar. These are the second and fourth icons from the right, respectively. The cursor will turn into a plus sign.

    - Click and drag the mouse to specify an area for the graph. For v. 4.0: The Chart Wizard will reappear with a series of dialog boxes that allow you to customize your graph.

        1st box: If the range of selected cells is correct, click 'Next'; Otherwise enter the correct range.
        2nd box: Select the 'Column' graph type (third choice, usually the default anyway). Click 'Next'.
        3rd box: Select the graph option number 8. Click 'Next'.
        4th box: Ignore this box, and simply click 'Next'.
        5th box: Add an appropriate title (e.g. 'Planetary spacing') and axis labels, then click 'OK'.
      For earlier versions of **Excel**: The graph will automatically appear as a bar graph (you can specify some other type by selecting options from the **GALLERY** menu). Double-click on the graph to customize it with a title and axis labels.

  > To print your document:

      Highlight the range of cells to be printed. You can include both the graph and your data, or just your graph. From the **OPTIONS** menu, select *Set print area*.

      - From the **FILE** menu, select *Print*. Click 'OK'.

  > Save the worksheet under the name 'PLANETS' to the appropriate drive and directory or folder.

NOTES FOR FUTURE USE:

<u>LAB 1/Exercise I.B.2., p. 4: Planetary spacing</u>

- **Lotus 1-2-3 (Release 4 for Windows):**
  - \> Open Lotus 1-2-3
  - \> To enter data:
    - In cells A1 and B1 respectively, type the words 'Planets' and 'Distance' (no quotes). These will act as titles for your columns of data.
    - In cells A2 to A10, enter the names of the planets, in order of increasing distance from the Sun.
    - In cells B2 to B10, enter the planets' distances from the Sun (Table 1-2) as logarithms. To do this, type '=@LOG(distance)' in each cell (no quotes). Numbers in scientific notation must be entered in modified form. For example, 5.79×10^7 would be 5.79E+07. So the full entry in B2 should be: '=@LOG(5.79E+07)'.
  - \> To create a graph :
    - First, highlight the range of cells to be graphed (cells A2 through B10).
    - Next, click on the chart button (second icon from right on the bar near the top of the screen)
    - Click and drag the mouse to specify an area for the graph.
    - When the graph is selected, the **CHART** menu appears. This menu allows you to change the graph type, headings, axis labels, etc. In this case, a bar graph (the default type) is fine.
    - To add a title to the graph, go to the **CHART** menu again and select *Headings*. Type an appropriate title (e.g. 'Planetary spacing') and click 'OK'.
    - To label and scale the axes, return to the **CHART** menu and select *Axis*. Enter appropriate labels and click 'OK'.
  - \> To print your document:
    - Highlight the range of cells to be printed. You can include both the graph and your data, or just your graph.
    - From the **FILE** menu, select *Print* . Click 'OK'.
  - \> Save the worksheet under the name 'PLANETS' to the appropriate drive and directory or folder.

NOTES FOR FUTURE USE:

LAB 1/ Exercise I.B.2., p. 4: Planetary spacing 

Borland Quattro Pro (v. 5.0 for Windows):

> Open Quattro Pro

> To enter data:

- In cells A1 and B1 respectively, type the words 'Planets' and 'Distance' (no quotes). These will act as titles for your columns of data.
- In cells A2 to A10, enter the names of the planets, in order of increasing distance from the Sun.
- In cells B2 to B10, enter the planets' distances from the Sun (Table 1-2) as logarithms. To do this, type '@LOG(distance)' in each cell (no quotes). Numbers in scientific notation must be entered in modified form. For example, 5.79×10^7 would be 5.79E+07. So the full entry in B2 should be: '@LOG(5.79E+07)'.

> To create a graph:

- From the **GRAPH** menu, select *New*.
- The "Graph New" window will appear. Give the graph a name (e.g. 'Planets').
- Next, click on the "X-Axis" box. The Graph New window will disappear. Select cells A2 to A10 (Mercury to Pluto) by clicking and dragging the mouse. This range will appear under the title of the Graph New window. Restore the window to full size by clicking the up button in the upper right.
- Click the "Legend" box and highlight the word 'Planets' in cell A1.
- Click on the "1st" box and highlight the logarithmic distance values in Column B.
- After returning to the Graph New window, click 'OK'. The plotted graph will appear. (To return to the spreadsheet, go to the **WINDOW** menu and select the name of the spreadsheet. If the worksheet hasn't been saved, the name will be NOTEBK.WB1).
- To give the graph a title, pull down the **GRAPH** menu and select *Titles*.

> To print your document:

- Highlight the block of cells to be printed. From the **FILE** menu, select *Print* . Click 'OK'.

> Save the worksheet under the name 'PLANETS' to the appropriate drive and directory or folder.

NOTES FOR FUTURE USE:

<u>LAB 1/ Exercise I.B.2., p. 4: Planetary spacing</u> 

- **Informix WINGZ (v. 1.1 for Windows; v. 1.1 for Macintosh):**

  > Open WINGZ

  > To enter data:

   - In cells A1 and A2, respectively, type the words 'Planets' and 'Distance' (no quotes). These will act as titles for your rows of data.

   - In cells B1 to J1, enter the names of the planets, in order of increasing distance from the Sun.

   - In cells B2 to J2, enter the planets' distances from the Sun (Table 1-2) as logarithms. To do this, type: '=LOG(distance)' in each cell (no quotes). Numbers in scientific notation must be entered in modified form. For example, 5.79×10^7 would be 5.79E+07. So the full entry in B2 should be: '=LOG(5.79E+07)'.

  > To create a graph:

   - Select cells A1 through J2 by clicking and dragging the mouse.

   - While these cells are highlighted, click once on the bar graph icon on the left side of the screen (fifth from top). Move the cursor to the left side of the worksheet beneath your data cells. Hold down the mouse button and drag down and right until you have outlined an area for the graph. A bar graph automatically appears. (The graph will be partly obscured if the area you outline is not large enough. If this happens, click on the lower right corner of the graph and drag it down and right).

   - The **GRAPH** menu contains many options for modifying your plot. (Note that the graph must be selected before the options under this menu are activated). For example, select *Legend* to change the position of the legend on your graph. In this case, you may want to remove it altogether (select *Hide legend*). You can add a title in the following manner. First, type an appropriate title (e.g. 'Planetary spacing') in an empty cell on your worksheet. Note that you must click on the open plus sign icon, top left side of screen, before entering additional text. Then select this cell and while holding down the control key (PCs) or Apple key (Macs), also select the graph. Then go to the **GRAPH** menu and select *Title*, then *Title range.*

  > To print your document:

   - Highlight the range of cells to be printed.

   - From the **SHEET** menu, choose *Report,* then *Report print range* .

   - From the **FILE** menu, select *Print* . Click 'OK'.

  > Save the worksheet under the name 'PLANETS' to the appropriate drive and directory or folder.

NOTES FOR FUTURE USE:

<u>LAB 1/ Exercise I.C.2., p. 6 (Planetary groupings)</u>: 

- **All programs:**
  - > To make plots of planetary masses, densities and diameters:
    - If the data files have already been created, open the file appropriate to your group.
    - If the files do not yet exist, enter the data appropriate to your group following the procedures you used to create the plot of planetary spacing (i.e., list the planets in order of distance from Sun and enter their corresponding masses, densities or diameters).
    - Create a bar graph illustrating the variations in planetary mass, density or diameter.
    - Print and save your plot if you are instructed to do so.

<u>LAB 1/ Exercise I.C.5., p. 7 (Adding the asteroids to the plot of planetary spacing)</u>

- **All programs:**

  Open the 'Planets' file you created earlier, then follow the instructions appropriate to your spreadsheet.

- **Microsoft Excel:**
  - > To add a row to the existing data:
    - Select the cell containing Mars. From the **EDIT** menu, select *Insert*. When prompted, indicate that a new row is to be added *after* the highlighted cell. Then click 'OK'.
    - Enter the asteroids and their average orbital distance (as a logarithm) in the new cells.
  - > Print the revised graph if you are instructed to do so.

- **Lotus 1-2-3:**
  - > To add a row to the existing data:
    - Select the cells containing Jupiter and its distance value. From the **EDIT** menu, select *Insert*. In the dialog box that appears, click on the "Row" option, then 'OK'.
    - Enter the asteroids and their average orbital distance (as a logarithm) in the new cells.
  - > Print the revised graph if you are instructed to do so.

- **Quattro Pro:**
  - > To add a row to the existing data:
    - Select the cells for Jupiter and its distance value. From the **BLOCK** menu, select *Insert,* then *Rows*.
    - Enter the asteroids and their average orbital distance (as a logarithm) in the new cells.
  - > Print the revised graph if you are instructed to do so.

- **WINGZ:**
  - > To add a column to the existing data:
    - Highlight the cells containing Jupiter and its distance value. From the **EDIT** menu, select *Insert*.
    - Enter the asteroids and their average orbital distance (as a logarithm) in the new cells.
  - > Print the revised graph if you are instructed to do so.

LABORATORY 2: *THE EARTH IN TIME* 

<u>LAB 2/Exercise III.A.6., p. 30: Tree ring exercise</u>

The goal of this exercise is to create a plot illustrating both yearly precipitation data and tree ring widths

- **All programs:**
  - > Open the spreadsheet program
  - > Open the existing file containing historical precipitation data for your area.
    - Calendar years should be listed in the first column; yearly precipitation values in the second.
  - > From the **FILE** menu, select *Save as* and give the worksheet a new name.
  - > Enter your tree ring widths in the third column. Make sure the values correspond to the right years.
  - > If one does not already exist, make a plot (line, x-y or scatter type) of the precipitation data (y-axis) vs. year (x-axis), following the steps learned in earlier exercises.
  - > Follow the instructions below to add your ring width data to the precipitation plot.
- **Excel:**
  - > Double-click on the graph. From the **CHART** menu, choose *Edit Series*. In the dialog box, chose 'New series'. Enter the range of cells of the tree ring data in the 'Y values' area. (You may have to enter the worksheet name in front of the cell range; follow the format in the 'X labels' box). Click 'OK'.
- **Lotus 1-2-3:**
  - > Double click on the graph. From the **CHART** menu, select *Ranges*. In the dialog box, click on the B data range line and enter the range of the tree ring data in the space near the bottom of the box. Click 'OK'. You can add a second scaled vertical axis by selecting *Axes*, then *2nd axis* from the **CHART** menu.
- **Quattro Pro:**
  - > From the **GRAPH** menu, select *Series*. In the dialog box, click the 'Add' button, then use the mouse to select the tree ring data. Click 'OK'.
- **WINGZ:**
  - > Select (click on) the symbol to the left of the word "Precipitation" in the legend, pull down the **GRAPH** menu and choose *Series,* then *Add series* .
  - > To add a data series to a WINGZ x-y or scatter plot, <u>two</u> series (columns) must be simultaneously selected while the new series symbol (here, for ring width) is also selected. Follow these instructions:
    - Select the new series symbol (if it's not already selected).
    - While holding down the control key (PCs) or Apple key (Macs), select the years column (but not the heading). With the control/Apple key still depressed, select the ring width data.
    - Finally, pull down the **GRAPH** menu and select *Series* then *Range.*
  - > Select the ring width series symbol again, then from the **FORMAT** menu, select *Line.* In the dialog box that appears, choose a new line width or color so that the two data series can be distinguished.
  - > With the ring width series symbol still selected, go to the **GRAPH** menu and select *Series* then *Aux axis.*

LAB 2/ Exercise III.B.5, p. 33: Changing ratios of daughter/parent isotopes

Spreadsheets are ideal for making repetitive calculations such as changes in daughter/parent isotope ratios.

- **All programs:**
    - > Set up a four-column table like that in exercise III.B.2., with the following in headings in cells A1 to D1:

        'Half-life #'   'Amount of parent'   'Amount of daughter'   'Daughter/parent ratio'

    - > In the first data row, enter values for the start of the decay process (Half-life = 0; Amount of parent = 1)

    - > In subsequent rows, enter appropriate formulas rather than numbers, according to instructions below.

        Symbols for basic math operations: Add: +   Multiply: *   Subtract: -   Divide: /   Raise to power: ^

- **Excel:** Simple arithmetic formulas in Excel begin with '='.
    - > Column A (Half-lives): In cell A3, type '=A2+1' (no quotes). Then select cell A3 and drag the mouse down to cell A8. Next, go to the **EDIT** menu and select *Fill Down.*
    - > Column B (Amount of parent): In cell B3, type '=B2*0.5', then copy this formula into cells B4-B8.
    - > Column C (Amount of daughter): In cell C3, enter a formula that gives the amount of daughter product after one half-life in terms of the original amount (C2) and amount of the parent (B3). Copy to C4-C8.
    - > Column D (Daughter/Parent ratio): Enter an appropriate formula in D3, then copy to D4-D8.

- **Lotus 1-2-3:** Simple arithmetic formulas in Lotus begin with '=+'.
    - > Column A (Half-lives): In cell A3, type '=+A2+1' (no quotes). Then select cell A3 and drag the mouse down to cell A8. Next, go to the **EDIT** menu and select *Copy Down.*
    - > Column B (Amount of parent): In cell B3, type '=+B2*0.5', then copy and paste this formula into cells B4-B8.
    - > Column C (Amount of daughter): In cell C3, enter a formula that gives the amount of daughter product after one half-life in terms of the original amount (C2) and amount of the parent (B3). Copy to C4-C8.
    - > Column D (Daughter/Parent ratio): Enter an appropriate formula in D3, then copy to D4-D8.

- **Quattro Pro:** Simple arithmetic formulas in Quattro Pro begin with '+'.
    - > Column A (Half-lives): In cell A3, type '+A2+1' (no quotes). With cell A3 selected, go to the **EDIT** menu and select *Copy* . Then *Paste* the formula into cells A4-A8.
    - > Column B (Amount of parent): In cell B3, type '+B2*0.5', then copy and paste this formula into cells B4-B8.
    - > Column C (Amount of daughter): In cell C3, enter a formula that gives the amount of daughter product after one half-life in terms of the original amount (C2) and amount of the parent (B3). Copy to C4-C8.
    - > Column D (Daughter/Parent ratio): Enter an appropriate formula in D3, then copy to D4-D8.

- **WINGZ:** Simple arithmetic formulas in WINGZ begin with '='.
    - > Column A (Half-lives): In cell A3, type '=A2+1' (no quotes). Then select cell A3 and drag the mouse down to cell A8. Next, go to the **EDIT** menu and select *Copy Down.*
    - > Column B (Amount of parent): In cell B3, type '=B2*0.5', then copy this formula into cells B4-B8.
    - > Column C (Amount of daughter): In cell C3, enter a formula that gives the amount of daughter product after one half-life in terms of the original amount (C2) and amount of the parent (B3). Copy to C4-C8.
    - > Column D (Daughter/Parent ratio): Enter an appropriate formula in D3, then copy to D4-D8.

LABORATORY 5: *EARTH'S AURA* 

<u>LAB 5/ Exercise I.B.2., p. 85: Quantifying the local hydrologic budget</u>

- **All programs:**
  - > Open the program and the file containing monthly precipitation, evaporation, and stream flow data.
  - > Total annual precipitation, evaporation and stream flow values can be determined easily from monthly data using the SUM formula. For example, if the precipitation data are in cells A2-A13, enter in A14:
    - Lotus 1-2-3: ' =@SUM(A2..A13)' (no quotes)   - Excel: '=SUM(A2..A13)' (no quotes)
    - Quattro Pro: '=@SUM(A2..A13)' (no quotes)   - WINGZ: '=SUM(A2..A13)' (no quotes)
  - > Note that the stream flow data must be treated somewhat differently since they are not volumes but volumes per unit time. As indicated in equation 5-4, sum the monthly flow rate values (using the SUM formula), then divide the total by 12 to get the average annual flow rate value. Multiply this by the number of seconds per year (3.16×10^7) to get the actual volume of water that flowed past the station.

LABORATORY 6: *THE ORGANIC EARTH* 

<u>LAB 6/ Exercise I.A.1., p. 104: Chemistry of the human body</u>

- **All programs:**
  - > In 5 columns with the same headings as in Table 6-1, enter the elements listed (column A) and their abundances in the human body (B), Earth's crust (C), atmosphere (D), and seawater (E). List 'Trace' amounts as 0.00. NB: In WINGZ, enter the data in *rows*, with the headings in cells A1 through A5.
- **Excel:**
  - > Select all cells containing entries (including the column headings) and use the **Chart Wizard** (v. 4.0) or Chart button (v. 3.0) to create a 3-D column graph. (In v. 4.0, select the chart type in the lower left corners of the second and third Chart Wizard dialog boxes).
- **Lotus 1-2-3:**
  - > From the **GRAPH** menu, select *New*
    - From the **CHART** menu, specify the chart type as 3-D Bar.
    - From the **CHART** menu, select *Ranges*. The X data range should be the elements in column A; the Data range should include all entries in columns B-E.
    - From the **CHART** menu, select *Legend*. Specify the column headings in B1..E1 as the legend range.
- **Quattro Pro:**
  - > From the **GRAPH** menu, select *New* and specify '3-D bar' as the graph type. Then pull down the **GRAPH** menu again, select *Titles* and add a title, legend, and axis labels for the graph.
- **WINGZ:**
  - > Select all cells containing entries (including the row headings), then click on the graph icon and define an area for the plot. You may leave the plot as it is, or select *3-D bar* from the **GALLERY** menu.

LABORATORY 7: *EARTH RESOURCES* 

Global lead consumption exercises (I.B.3.-5., pp. 128-131)

In Laboratory 7 of this manual we explore several models of lead consumption by performing calculations based on current consumption, estimated reserves, and changing rates of consumption. Such modeling is easily carried out on spreadsheets. The instructions below follow the questions on pp. 128-131 in Lab 7, where we start with "known" lead reserves of 120,000,000 tons in 1989, and consumption of 6,000,000 tons that year. We will first create a "static model", in which lead consumption remains constant; then an "exponential model" in which consumption increases by a certain rate each year; and finally a model in which the initial "known" reserves are underestimated by a factor of 5.

Static Model (Exercise I.B.3., p. 128):

- **All programs:**
    > Open the program and a blank worksheet.
    > In cell A1, type 'Year' (no quotes).
      - In cell A2, type '1989', the year for which you have initial data.
      - In cells A3-A33, enter calendar years 1990-2020 by typing an appropriate formula in A3 and copying it to the cells below (review instructions for daughter/parent isotope exercise, p. 169).
    > In cell B1, and type 'Lead Consumed (tons)'.
      - In cell B2 type '6000000' (no quotes). This represents the 6,000,000 tons of lead consumed in 1989. Copy the contents of this cell to cells B3 through B33. (In this model, the same amount of lead is consumed each year).
    > In cell C1 type 'Available Lead (tons)'. This column will list the lead available at the end of each year.
      - At the end of 1989, the estimated available lead was 120,000,000 tons. Enter this value in cell C2.
      - To calculate the lead available in subsequent years, we need to create a formula. The lead available in 1990 will be the starting amount minus that used in 1989. In cell C3, type '=C2-B3' (Excel and WINGZ); '=+C2-B3' (Lotus); or '+C2-B3' (Quattro Pro). Then copy this formula into cells C4–C33.

Exponential Model (Exercise I.B.4., pp. 129-130):

To apply an exponential model of lead consumption, we need only to modify the 'Lead Consumed' column.

- **All programs:**
    > Select cells B3-B33 and delete their contents.
    > For the new model, begin with the assumption that the rate of consumption increases by 3%/year.
      - To make the spreadsheet as flexible as possible, enter this value as a decimal (0.03) in cell D1.
      - In cell B3 type the formula '=B2*(1+$D$1)' (Excel and WINGZ); '=+B2*(1+$D$1)' (Lotus); or '+B2*(1+$D$1)' (Quattro Pro). Then copy the formula to cells B4-B33. The $ signs are 'absolute cell references', indicating that you want the value in cell D1 to be used for all of column B.
    > To change the growth rate of consumption of a resource, simply change the value in cell D1.

Changing estimates of remaining reserves (Exercise I.B.5., pp. 131):

- **All programs:**
    > Change the entry in cell C2, the initial value in the 'Lead available' column, to a value of 600,000,000 tons (5 times the present estimate of reserves).

APPENDIX II: METRIC MEASURES AND CONVERSION FACTORS

The metric system, or Systeme Internationale (SI), is a system of measures based on factors of 10. Because of its mathematical simplicity, the metric system is now used for commercial purposes by almost all countries in the world -- with the notable exception of the United States. Even in the US, however, the metric system is used in science. Metric measures are generally used in the laboratory exercises, except in a few cases where maps you will use are printed in English units. The following information is provided for the benefit of those who still think in bushels, ounces and furlongs.

A. Basic units in the metric system

Length:	meter (m)	*Volume:*	liter (l)
Mass:	gram (g)	*Temperature:*	°Celsius (°C)

B. Prefixes in the metric system

micro-:	10^{-6} (millionths)	kilo-:	10^3 (thousands)
milli-:	10^{-3} (thousandths)	mega-:	10^6 (millions)
centi-:	10^{-2} (hundredths)	giga-:	10^9 (billions)

C. Metric-English conversion factors

Quantity	Unit	Multiply by*	To get
Length	meter	3.28	feet
	centimeter (cm)	0.3937	inches
	kilometer (km)	0.6214	miles
Mass	gram (g)	0.03527	ounces
	kilogram (kg)	2.205	pounds
Volume	liter (l)	0.946	quarts

Also: 1 milliliter (ml) = 1 cm^3

1 cm^3 of water at room temperature has a mass of 1 g

Quantity	Unit	Multiply by*	To get
Temper-ature	°C	$\times 9/5 + 32$	°F
	°F	$- 32 \times 5/9$	°C

* To convert from English system, <u>divide</u> by the conversion factor.

APPENDIX III: UNDERSTANDING TOPOGRAPHIC & PHYSIOGRAPHIC MAPS

- Both topographic and physiographic maps portray three-dimensional landscape features on two-dimensional surfaces (sheets of paper).

- Physiographic maps use colors, shading or patterns to depict landforms in a generalized way.

- Topographic maps provide more detailed information about elevation because they have <u>contour lines</u> -- lines connecting points of equal elevation on the landscape. If you've never used a topographic map before, you will find these basic principles useful:

> Imagine yourself walking along the side of a steep hill, trying to stay at the same elevation. You'd be tracing out the path of a contour line.

> The vertical distance between adjacent contour lines is the <u>contour interval</u>. The contour interval on a particular map depends on the scale of the map and the ruggedness of the topography. For example, a 5-foot contour interval might be appropriate for a local map of the flat landscape of North Dakota, but would be impractical on a regional map of the Rockies.

> Contour lines are usually labeled with the elevation they represent, though on some maps only every fourth or fifth line is labeled. By knowing the contour interval, you can determine the elevation of the unlabeled lines.

> Closely-spaced contours indicate steep slopes; wide spacing indicates flatter areas.

> Contour lines can never cross or divide, nor end in the middle of a map; they either extend off the map or form closed loops (indicating hills or depressions):

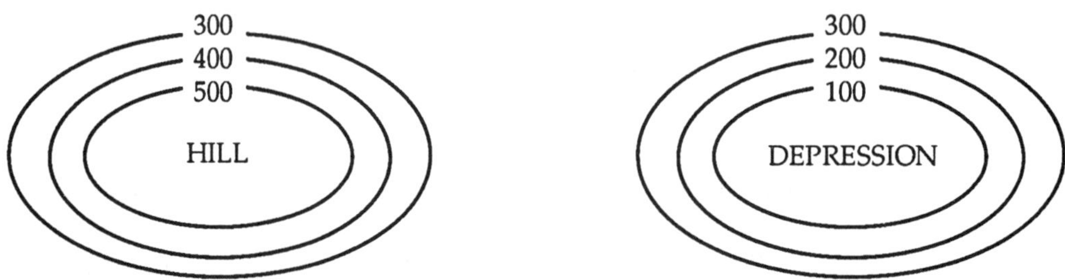

> Stream valleys can be recognized by V-shaped contour lines that close upstream:

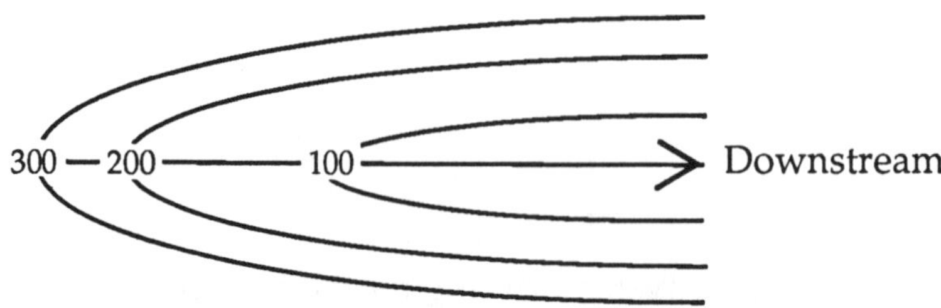

APPENDIX IV: ROCK AND MINERAL IDENTIFICATION CHARTS

A. MINERAL PROPERTIES

Table IV-2 (next page) summarizes the characteristics of 20 major rock-forming minerals (and mineral groups) featured in exercises in the lab manual. The columns list the properties that are most useful in hand specimen identification of minerals, including:

- **Crystal form:** Among the most diagnostic properties because it reflects the atomic-scale structure of minerals. Quartz crystals in geodes, for example, are unmistakable because of their 6-fold symmetry. However, because most mineral specimens did not grow in unconfined spaces, they do not show their full crystal form. In such cases, **cleavage** is the next property to examine.

- **Cleavage** (tendency to break along surfaces of particular orientations): Cleavage surfaces represent planes of relatively weak bonds within the mineral lattice and are therefore very reliable criteria for mineral identification. For example, micas always cleave into thin sheets, while calcite breaks consistently along 3 oblique planes to create rhombohedral blocks.

- **Hardness:** Also a reflection of atomic-scale structure, hardness is usually described in terms of relative values on <u>Moh's hardness scale</u>. The scale ranks 10 "index" minerals according to their resistance to being scratched. A mineral high on the scale will scratch all those below it:

TABLE IV-1: MOH'S HARDNESS SCALE

<u>Hardness value</u>	<u>"Index" mineral</u>	<u>Practical uses</u>	<u>Objects of same hardness</u>
10 (hardest)	Diamond	Drill bits	
9	Corundum	Abrasives	
8	Topaz	Gemstone	
7	Quartz	Watches	
6	Potassium feldspar		Pocketknife, glass
5	Apatite	(Tooth enamel)	
4	Fluorite	Flux in steel-making	
3	Calcite	Cement	Copper penny
2	Gypsum	Plaster	Fingernail
1 (softest)	Talc	Talcum powder	

- **Density** (mass per unit volume): Very diagnostic, but accurate determination requires careful measurement of specimen mass and volume. The high density of some minerals (e.g. the lead ore galena) is detectable simply by holding a specimen.

- **Luster** (character of light reflected from mineral): Most useful for identifying metallic minerals.

- **Streak** (color produced when mineral is drawn across a ceramic plate): Very diagnostic for some minerals. Streak color may be quite different than the color of the specimen itself; for example sphalerite (zinc sulfide) is typically brownish in appearance but leaves a yellow streak.

- **Color:** Not a reliable criterion for mineral identification because trace amounts of impurities can dramatically change the color of a mineral. Quartz, for example, can be clear, smoky, rose-tinted or purple (amethyst).

- **Other:** Some mineral have distinctive properties that make them easy to recognize. For example, halite tastes salty; magnetite is magnetic; calcite effervesces when hydrochloric acid is placed on it.

When crystals are too small for hand specimen identification, minerals may be identified by examining transparently thin sections of them under the microscope. The atomic structure of minerals determines the manner in which light passes through them, giving each distinctive optical properties.

TABLE IV-2: PROPERTIES OF MAJOR MINERALS

Adapted with permission from Skinner, B. & S. Porter, 1992. *The Dynamic Earth: An Introduction to Physical Geology.* New York: Wiley. (Appendix C)

Mineral (page cited in text)	Chemical formula	Nature & occurrence	Crystal form	Cleavage	Hardness/ Density, g/cm^3	Color, luster, streak, other	Most distinctive properties
*Amphiboles** (9, 50) - Hornblende - Actinolite - Tremolite	$X_2Y_5Si_8O_{22}(OH)_2$ where X= Ca, Na Y= Mg, Fe Al	Igneous and metamorphic rocks	Long 6-sided crystals. May be fibrous	2 planes intersecting at 56° and 124°	H:5-6 D: 2.9-3.8	Hornblende is dark green to black; Actinolite green; Tremolite white	Crystal form, cleavage
Apatite (108)	$Ca_5(PO_4)_3(F,OH)$	Igneous rocks; main source of P for biosphere	6-sided crystals.	Poor, one direction	H: 5 D: 3.2	Green, brown, blue or white	Hardness, crystal form
Calcite (78, 91)	$CaCO_3$	Chemical sedimentary rocks	Tapering crystals	3 oblique; form rhombohedral fragments	H: 3 D: 2.7	Clear or white Effervesces with HCl	Cleavage, hardness, effervescence
Chalcopyrite (129)	$CuFeS_2$	Copper ore	Massive	None	H: 3.5-4 D: 4.2	Golden to brassy	Dark green streak
Dolomite (91)	$CaMg(CO_3)_2$	Chemical sedimentary rocks	Rhomb-shaped	3 directions as in calcite	H: 3.5 D: 2.8	White to grey Does not fizz in HCl unless scratched	Cleavage, Lack of strong effervescence
Feldspar (13,41,125) - Plagioclase (Anorthite/Albite)	$Ca\ Al_2Si_2O_8$ $NaAlSi_3O_8$		Tabular	2, not quite at right angles	H: 6-6.5 D: 2.6-2.7	White to dark grey	Cleavage, fine lines on clv planes
- Orthoclase	$KAlSi_3O_8$	Igneous, metamorphic rocks	Prism-shaped	2 at right angles	H: 6 D: 2.6	Pink, white or grey	Color, cleavage
Galena (129)	PbS	Lead ore	Cubic crystals	3 at right angles	H: 2.5 D: 7.6	Grey;Dk grey streak	Form, density
Garnet (50, 125)	$X_3Y_2(SiO_4)_3$ X=Ca, Mg, Fe, Mn Y=Al, Fe, Ti, Cr	Metamorphic rocks	12 or 24-sided "soccer-ball" crystals	None; fractures unevenly	H: 6.5-7.5 D: 3.5-4.3	Red, brown, yellow-green or black	Form, hardness, lack of cleavage
Graphite (91)	C	Metasediments	Scaly, flaky	1 direction	H: 1-2 D: 2.2	Grey; metallic	Writes on paper

* Mineral names in italics identify major mineral families or groups with species of varying composition but common crystal structure.

TABLE IV-2, continued: PROPERTIES OF MAJOR MINERALS

Mineral (page cited in text)	Chemical formula	Nature & occurrence	Crystal form	Cleavage	Hardness/ Density, g/cm^3	Color, luster, streak, other	Most distinctive properties
Gypsum (78)	$CaSO_4 \cdot 2H_2O$	Evaporites	Long, tabular	1 direction	H: 2 D: 2.3	Clear; vitreous	Hardness
Halite (78)	$NaCl$	Evaporites	Cubic	3 at right angles	H: 2.5 D: 2.2	Clear to blue	Tastes salty
Hematite (115)	Fe_2O_3	Banded iron formations	Massive	None; fractures unevenly	H: 5-6 D: 5	Red- brown, grey or black Red-brown streak	Streak, hardness
Kaolinite (125)	$Al_2Si_2O_5(OH)_4$	Weathering product of feldspar; used in papermaking	Massive	1 direction	H: 2-2.5 D: 2.6	White to yellowish Sticks to tongue Smells clayey	Texture
Magnetite (115)	Fe_3O_4	Igneous, meta- morphic rocks; banded iron fms.	Octahedral crystals	None; fractures unevenly	H; 5.5-6.5 D: 5	Black; black streak Magnetic	Streak, magnetism
Micas (41, 50, 125) Biotite	$K(Mg, Fe)_3$- $AlSi_3O_{10}(OH)_2$	Igneous, meta- morphic rocks	Flakes, sheets	1 direction, cleaves into thin sheets	H: 2.5-3; D: 2.8-3.2	Black, brown, dark green	Cleavage, color. Flakes are flexible
Muscovite	KAl_3- $AlSi_3O_{10}(OH)_2$				H: 2-2.5; D: 2.7	Clear, white, pale green or brown	
Olivine (10)	$(Mg, Fe)_2SiO_4$	Igneous rocks	Massive	None	H: 6.5-7 D: 3.2-4.3	Olive, yellow-green	Color
Pyroxenes (13, 41)	$XY(SiO_3)_2$ $X=Y=Ca, Mg, Fe$	Igneous, meta- morphic rocks	8-sided, stubby crystals	2 at nearly right angles	H: 5-6 D: 3.2-3.9	Augite is dark green to balck; Others white to green	Cleavage
Quartz (41, 78, 96)	SiO_2	Igneous, meta- morphic and sedimentary rocks	6-sided prismatic crystals	None	H: 7 D: 2.6	Clear, white, smoky, pink, purple	Form; fine lines lines across faces at right angles to long dimension

B. ROCK PROPERTIES

Although rocks are composed of minerals, it is not always easy to identify individual minerals within them. The aggregate properties of rocks can also be used to recognize and classify them. Equally important is the geologic context in which rocks occur (one wouldn't, for example, expect to find an isolated layer of metamorphic gneiss in the middle of a sedimentary sequence). Rocks in the laboratory, have been removed from their 'natural habitats', and so hand specimen properties must be used in their identification.

Igneous, sedimentary or metamorphic?

The first step in identifying a rock specimen is to determine which of the three major rock groups it belongs to. Because rock types within the three catagories can have such a wide range of properties, this is not always easy. Use the criteria below as a guide.

If the specimen has <u>2 or more</u> of the following characteristics, it is probably **igneous**:
- Interlocking crystals
- "Foamy" texture with lots of open spaces
- Glassy texture
- Relatively hard, compact

If the specimen has <u>2 or more</u> of the following characteristics, it is probably **sedimentary**:
- Bedding and/or other sedimentary structures (ripples, mudcracks)
- Fossils
- Rounded grains
- Relatively soft, crumbly

If the specimen has <u>2 or more</u> of the following characteristics, it is probably **metamorphic**:
- Strongly aligned crystals
- Shiny micaceous surfaces
- Streaky layering
- Relatively hard, compact
- Visibly distorted bedding, fossils, etc.

Once the first-order categorization has been determined, the rock's identity can be further constrained using classification keys like those in Figures IV-1, IV-2 and IV-3.

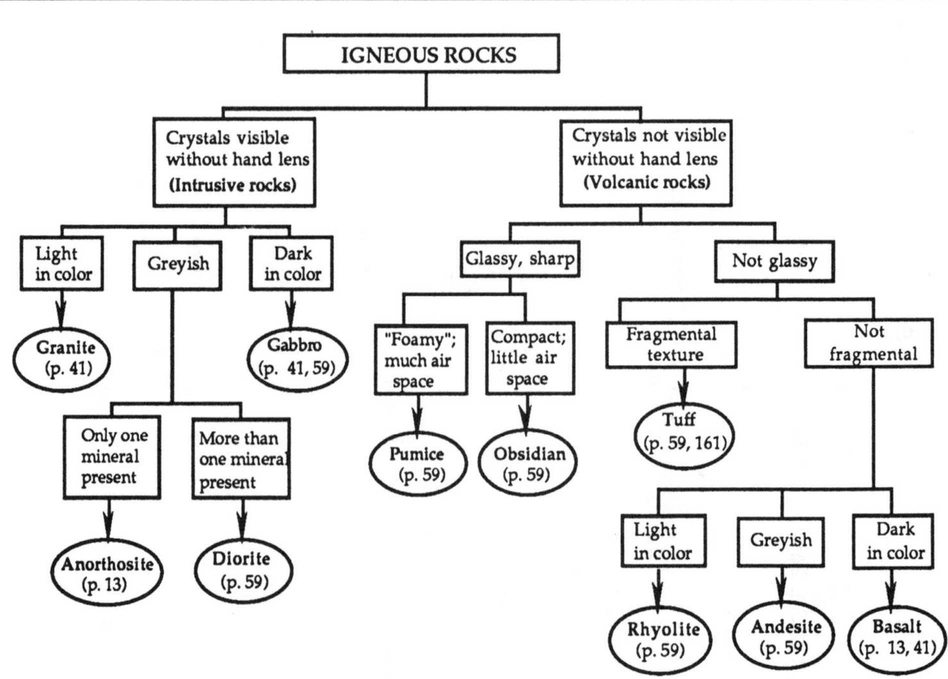

FIGURE IV-1.
CLASSIFICATION
KEY FOR
IGNEOUS ROCKS

FIGURE IV-1.
CLASSIFICATION KEY FOR SEDIMENTARY ROCKS

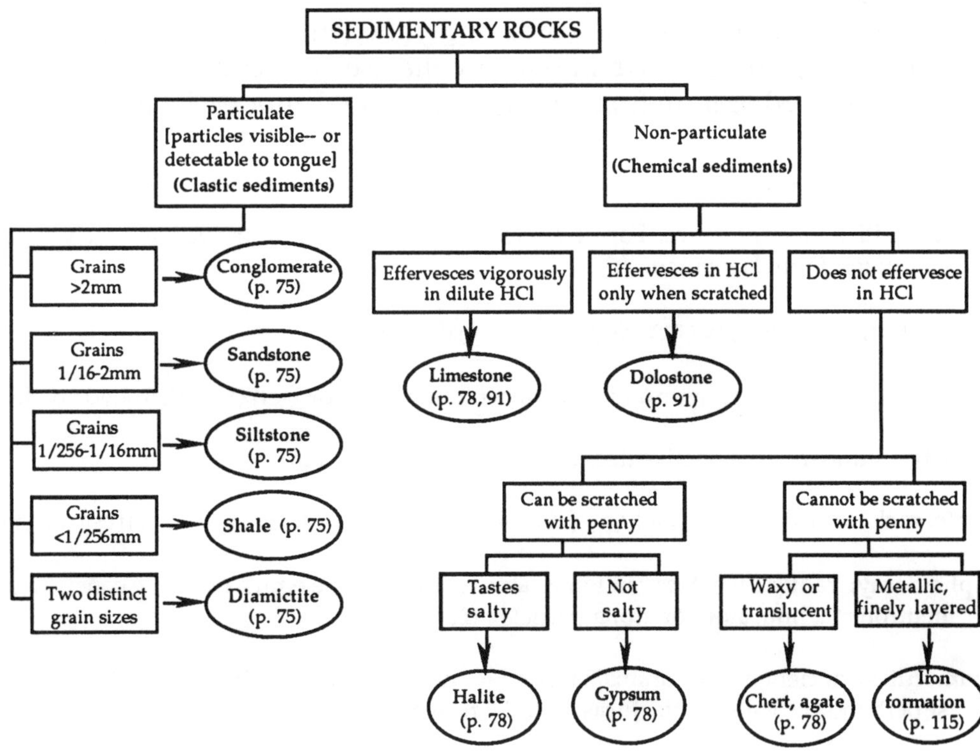

FIGURE IV-1.
CLASSIFICATION KEY FOR METAMORPHIC ROCKS

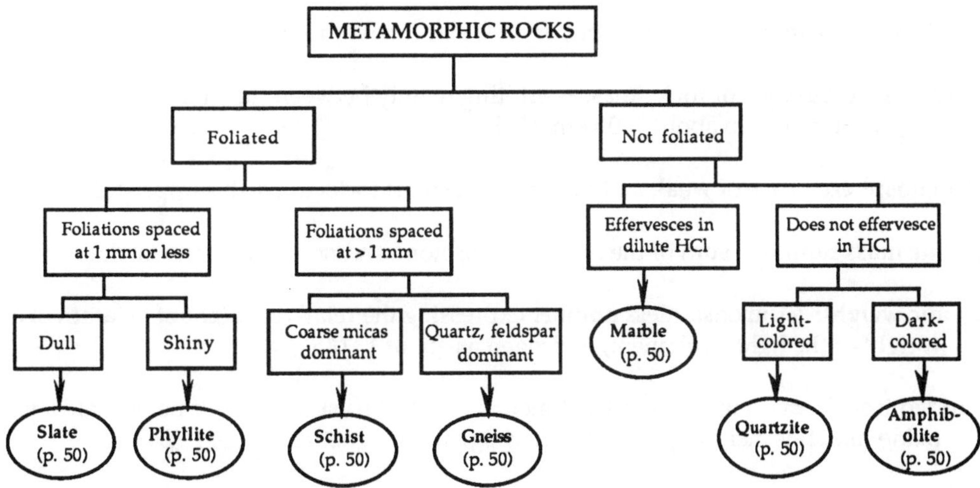

APPENDIX V: GLOSSARY

The following list includes all words <u>underlined</u> in the text of the manual. In the definitions, words shown in **bold** are themselves defined elsewhere. The laboratory in which a word or term first appears is indicated after each entry (e.g. *L2 = Laboratory 2, Earth in Time*).

---

aa: lava with a blocky texture. Formed when the surface of basaltic lava has nearly solidified but flow continues. From Hawaiian. *(L3)*

absolute age: actual age of a geologic event in years. *(L2)*

acidic solution: solution with a higher concentration of hydrogen (H^+) than hydroxide (OH^-) ions. *(L5)*

activity: number of radioactive decay events per unit time. *(L8)*

agate: rock formed by precipitation of dissolved silica (SiO_2) from seawater, freshwater or groundwater. *(L4)*

alpha decay: radioactive decay in which an unstable isotope emits a particle consisting of 2 protons and 2 neutrons (helium atom or alpha particle). *(L8)*

ammonites: ancient marine animals related to modern squid. Their detailed shell ornamentation makes them superb index fossils for the Jurassic and Cretaceous periods. *(L2)*

andesite: volcanic rock with composition intermediate between that of basalt and rhyolite. Most common rock type at convergent plate boundaries. Intrusive equivalent: diorite. *(L3)*

anorthosite: a slowly-cooled igneous rock composed mainly of the mineral anorthite. *(L1)*

anthracite: high-grade coal. *(L7)*

aquifer: permeable layer of sediment or rock that carries groundwater. *(L4)*

asteroids: thousands of rocky masses orbiting mainly between the inner and outer planets, ranging in size from centimeters to about 1000 km. *(L1)*

asthenosphere: zone of weak and flowing (but solid) rocks in Earth's upper mantle. *(L3)*

atomic mass number: sum of the number of protons and neutrons in the nucleus of an atom. *(L2, L8)*

atomic weight: a dimensionless number expressing the relative weight of an atom of an element. The atomic weight of the lightest element, hydrogen, is about 1. *(L5)*

banded iron formation: chemical sedimentary rock consisting of centimeter-scale layers of **chert** and iron oxide minerals including hematite (Fe_2O_3) and magnetite (Fe_3O_4). *(L6)*

basalt: low-silica (mafic) volcanic rock, consisting mainly of Ca- and Na-rich feldspar and pyroxenes. Typical of divergent plate boundaries and oceanic hot spots (and also the lunar lowlands). Intrusive equivalent: gabbro. *(L1, L3)*

base level: elevation of the body of water into which a river flows. *(L4)*

bedding: depositional layering in a sedimentary rock. *(L4)*

beta decay: radioactive decay in which unstable isotopes emit electrons from the nucleus (beta particles). *(L8)*

bituminous coal: medium-grade coal. *(L7)*

Bode's rule: mathematical description of planetary spacing. For you to define! *(L1)*

brittle deformation: change in shape by fracturing and frictional sliding. *(L3)*

Carboniferous period: interval of geologic time, 360-285 million yrs before present. A time of global coal formation. In North America, the period is subdivided into the Mississippian and Pennsylvanian periods. *(L7)*

chemical sediment: sediment formed by chemical precipitation of minerals dissolved in water. *(L4)*

chemical weathering: decomposition of rocks by chemical processes. *(L5)*

chert: rock formed by precipitation of dissolved silica (SiO_2) from seawater, freshwater or groundwater. *(L4)*

clastic sediment: sediment deposited physically by water, wind or ice. *(L4)*

compound: a chemically bound combination of elements. Water is a compound of the elements hydrogen (H) and oxygen (O). *(L1)*

conglomerate: rock composed of pebbles and larger grains (>2 mm); deposited by fast-flowing water. *(L4)*

continental shelves: areas of shallowly submerged continental crust bordering the continents. *(L3)*

contour line: line joining points on a map where some quantity (e.g. elevation, barometric pressure) has the same value. *(L8)*

convection (thermal): gravitationally driven overturn of a heated material. *(L3)*

convergent plate boundary: boundary at which plates move together via **subduction.** *(L3)*

core: innermost and densest part of a planet. On Earth, the metallic core begins at a depth of 2900 km. *(L1)*

correlation: the process of relating rock layers in one area to rocks of the same age in another area. *(L2)*

cross bedding: layering that lies at an angle to overall bedding in a rock, representing the downstream faces of ripples or dunes. *(L4)*

cross-cutting relationships (principle of): principle that if feature A is cut or modified by feature B, then A must be older. *(L2)*

cross section: schematic drawing of a vertical cut into the subsurface. *(L2)*

crust: outermost layer of Earth, 5-90 km thick. *(L2)*

delta: sediment deposited at the mouth of a river as it enters a standing body of water. Often triangular in shape, resembling the Greek letter delta. *(L4)*

dendritic: branching or tree-like in form. *(L4)*

denitrification: return of fixed nitrogen -- NH_4^+, NH_3^+ (ammonia), and NO_3^- (nitrate) -- to the atmosphere as N_2. *(L6)*

density: mass of an object per unit volume. *(L1)*

desiccation cracks (mudcracks): polygonal cracks formed on the surface of fine-grained sediment as it dries and contracts. *(L4)*

diabase: intrusive igneous rock with composition similar to that of basalt. *(L2)*

diamictite: rock composed of two distinct sizes of clastic grains; diamictites usually represent glacial deposits or mud flows. *(L4)*

divergent plate boundary: boundary at which plates move apart. In an ocean basin, a *mid-ocean ridge. (L3)*

divide: topographic high defining the boundary of a **drainage basin.** *(L4)*

drainage basin: the land area drained by a particular river or river system. *(L4)*

ductile deformation: change in shape by flow rather than fracturing. *(L3)*

ejecta: rock material thrown out explosively from the site of an impact. *(L1)*

element: one of the fundamental species of matter. *(L1)*

epicenter: point on the surface of the earth above the **focus** of an earthquake. *(L3)*

equipotential lines: lines connecting points on a map where the water table is at the same elevation. *(L8)*

estuary: bay formed when rising sea level submerges the lower reaches of a river. *(L4)*

evaporite: rock composed of salts including halite, gypsum and anhydrite, formed by evaporation of saltwater. *(L4)*

exponential growth: growth of some quantity by amounts that depend on the existing amount. *(L7)*

exothermic reaction: chemical reaction that gives off heat. *(L7)*

extrusive rock: volcanic igneous rock, erupted onto Earth's surface. *(L3)*

fault: rock fracture along which slip has occurred. *(L2)*

feedback mechanism: a process in which the output (effect) of one phase becomes the input (cause) of the next. *(L5)*

flux: process that moves a material from one reservoir to another at a particular rate. *(L5)*

focus (earthquake): the origin of an earthquake; point in the subsurface where fault slip begins. *(L3)*

fold: wrinkle or buckle in layered rock, usually formed in response to layer-parallel shortening. *(L3)*

foliation: planar parting surface in a rock, formed by alignment of platy minerals like micas. *(L3)*

food chain: a hierarchy of organisms in which energy and nutrients are transferred from primary producers to the organisms that consume them, to the organisms that consume those organisms, etc. *(L6)*

fossil: preserved remains or traces of a living organism. *(L2)*

fractal: geometric feature that looks the same over a range of scales. *(L4)*

fractionation: separation of a material into components with distinct properties. Opposite of mixing. *(L1)*

Gaia hypothesis: view that Earth's near-surface environment has been largely controlled by biological activity -- i.e. that organisms have acted to minimize fluctuations in surface temperatures and air and water chemistry. *(L6)*

galena: lead sulfide (PbS); principal **ore** of lead. *(L7)*

gamma decay: radioactive decay in which an unstable isotope emits a form of electromagnetic radiation called a gamma ray. *(L8)*

gauging station: station on a river where water height and flow rate are monitored. *(L5)*

geomagnetic time scale: time scale based on the history of reversals in the polarity of Earth's magnetic field. Absolute ages have been assigned to the reversals through isotopic dating of rocks formed close to the time of a reversal. *(L3)*

geothermal gradient: increase in temperature with depth in Earth's subsurface, expressed as °C/km. *(L5)*

gneiss: intensely metamorphosed rock with prominent mineral banding. *(L2)*

gradient: slope of a river channel, expressed in feet per mile, meters per kilometer, or as a percentage. *(L4)*

granite: intrusive igneous rock intruded into subsurface, consisting principally of sodium/potassium feldspar, quartz, mica. *(L2)*

graptolites: possible predecessors of earliest vertebrates; excellent Silurian index fossils. *(L2)*

gravitational differentiation: settling of materials according to density. *(L1)*

greenhouse effect: trapping of heat close to a planet's surface by gases that transmit incoming sunlight but block heat reflected back from Earth's surface. Carbon dioxide (CO_2), Methane (CH_4) and water vapor (H_2O) can all act as greenhouse gases. *(L1)*

groundwater: water contained in the pore spaces of soil, sediment and rock. *(L4)*

half life: time required for half the amount of an unstable parent isotope to decay to a daughter isotope. *(L2)*

highlands (lunar): older, heavily cratered terrain on the Moon. *(L1)*

histogram: plot showing the frequency with which particular values of a variable occur. *(L1)*

hot spot: point on Earth's surface above an isolated column of unusually hot, buoyantly rising mantle rock. Magma generated as this rock nears the surface causes volcanic activity like that in Hawaii. *(L3)*

hydrocarbons: compounds of hydrogen and carbon, many of which can be burned as fuels (e.g. benzene, methane). *(L7)*

hydrologic budget equation: equation that accounts for all water entering or leaving an area. *(L5)*

hydrologic cycle: the continuous cycling of water in different forms at, above and under Earth's surface. *(L5)*

igneous rock: rock formed from the molten state (magma). *(L1)*

inclusion (principle of): principle that a rock containing pieces of other rocks must be younger than those pieces (or conversely, that the pieces must be older than the rock that contains them). *(L2)*

index fossil: fossil of an organism whose species existed for a geologically brief time period. Index fossils are very useful in correlation. *(L2)*

intrusive rock: igneous rock emplaced into the subsurface. *(L3)*

iridium: element 77, very rare on Earth, but found in higher concentrations in meteorites. High iridium concentrations in rocks from the Cretaceous-Tertiary boundary first lead to speculations that a meteorite impact may have lead to the extinction of the dinosaurs. *(L1)*

isochron (magnetic): line on an ocean floor map joining rocks formed during a particular magnetic reversal. From Greek: iso = same + kronos = time. *(L3)*

isostasy: the state of flotational balance of the crust and uppermost mantle on the asthenosphere. *(L3)*

isotope: atom of an element that can have varying numbers of neutrons in its nucleus. *(L2)*

joint: fracture in rock along which no slip has occurred. *(L3)*

lignite: low-grade coal. *(L7)*

limestone: rock composed mainly of calcite; typically precipitated from seawater. *(L4)*

limiting nutrient: element or molecule essential for life but available in limited quantities; if more were available, an organism's growth would be enhanced. *(L6)*

linear growth: growth of some quantity by constant amounts in a given period of time. *(L7)*

lithosphere: strong outer layer of the solid Earth, including the crust and uppermost mantle. Typically about 100 km thick. *(L3)*

logarithm: exponent to which a base number must be raised to produce a given number. For example, the logarithm of 100 in base 10 is 2. *(L1)*

logarithmic scale: measurement scale in which each unit represents an order-of-magnitude difference in value. The Richter scale is a logarithmic scale. *(L3)*

lowlands (lunar): younger, lava-covered terrain on the Moon. *(L1)*

mafic lava: low-viscosity, low silica lava. Solidifies to form basalt. *(L3)*

magnitude (earthquake): quantitative description of the size of an earthquake, based on amplitude of arrivals on seismic records. *(L3)*

mantle: layer surrounding core of the Inner planets. Earth's rocky mantle constitutes 83% of planet's volume and lies between about 40 km and 2900 km. *(L1)*

mass extinction: abrupt disappearance of many species from the fossil record. *(L2)*

metamorphic rock: rock modified by high temperature and/or pressure. *(L2)*

metamorphism: growth of new minerals in rocks in response to high pressure and/or temperature. *(L3)*

Milankovitch orbital cycles: cyclical variations in Earth's orbit and rotation that affect the amount of solar energy the planet receives and may influence climate. *(L5)*

mineral: a natural crystalline solid (one in which atoms are arranged in regular three-dimensional arrays) with a well-defined chemical composition. Most rocks consist of more than one type of mineral. *(L1)*

mode (statistical): the most common value of or range of values of a variable on a **histogram**. *(L3)*

model: simplified representation of a system. *(L6)*

mutation (genetic): a change in the genetic code (DNA) within a cell. Some mutations are harmless; others can lead to cancer and birth defects. *(L8)*

natural selection: evolution of organisms by the selective "survival of the fittest". *(L6)*

negative feedback mechanism: a self-correcting process. *(L5)*

nitrogen fixation: biochemical process in which bacteria convert atmospheric nitrogen (N_2) to compounds including ammonium (NH_4^+). *(L6)*

noble gas: any of the gaseous elements that occur in the last column of the periodic table. Having full outer electron shells, they do not bond with other elements. *(L8)*

non-renewable resource: resource that cannot be regenerated on a human time scale. *(L7)*

obsidian: volcanic glass. *(L3)*

ore: economically viable source of a metallic element. *(L7)*

oxidation (of an element/compound): ionization by loss of electrons, usually to oxygen atoms. *(L6)*

paleogeographic map: map showing configurations of oceans and land masses at times in the past. *(L3)*

pahoehoe: lava with a "ropy" surface texture. Typically formed in thin basalt flows. From Hawaiian. *(L3)*

permeability: measure of the ease with which fluids can move through sediment or rock. *(L4)*

phase diagram: plot showing the physical conditions under which the different states (solid, liquid, or gas forms) of a substance exist. *(L5)*

phosphorite: chemical sedimentary rock made of the mineral carbonate apatite $Ca_5(PO_4,OH,CO_3)_3(F)$, typically deposited in low latitude, shallow marine environments where upwelling currents bring deep-ocean phosphorous to surface waters. *(L6)*

photosynthesis: process by which green plants, with sunlight as their energy source, convert water and atmospheric carbon dioxide to sugars and free oxygen. *(L6)*

plate tectonics: theory that Earth's lithosphere is a mosaic of moving pieces or plates. *(L3)*

plume: area of subsurface groundwater contamination that originates from an identifiable source. *(L8)*

polarity (magnetic): the sense of magnetization of a body -- i.e., the relative positions of its north and south poles. *(L3)*

point source: location that can be considered the origin of a spreading pollutant. *(L8)*

porosity: amount (percent) of pore space in a sediment or rock. *(L4)*

porphyroblast: large crystal formed during metamorphism. *(L3)*

porphyry: igneous rock with two sizes of crystals: large ones formed slowly in a chamber under a volcano and finer ones formed by rapid cooling of lava erupted at the surface. *(L3)*

positive feedback mechanism: a self-perpetuating process. *(L5)*

primary producers: organisms able to derive energy from inorganic sources. Also called *autotrophs*. *(L6)*

prokaryotes: single-celled organisms, including algae and bacteria, that lack nuclei. *(L6)*

profile (of a river): cross-sectional view of a river channel in the downstream direction. *(L4)*

protolith: original rock from which a **metamorphic** rock formed. *(L2)*

pumice: rock formed from gas-rich, glassy volcanic "foam". Pumice may have so much air space that it will float! *(L3)*

pyroclastic material: volcanic debris thrown into the air during an explosive eruption; ranges in size from fine dust (ash) to >10 m blocks (bombs). Also called *tephra*. *(L3)*

radioactive decay or **radioactivity:** spontaneous break-down of unstable **isotopes** into heat, subatomic particles and other isotopes. *(L2)*

ratio: the quotient of two numbers. *(L2)*

red beds: clastic sedimentary rock deposited on land and colored red-orange by oxidized iron. *(L6)*

reduction (of an element/compound): ionization by addition of electrons. *(L6)*

remediation: clean-up and restoration of a polluted site. *(L8)*

renewable resource: resource that cannot be regenerated on a human time scale. *(L7)*

reservoir: temporary "storage place" for a particular material. *(L5)*

residence time: average length of time a material remains in a particular setting. *(L3, L5)*

rhyolite: high-silica (silicic) volcanic rock consisting mainly of K- and Na-rich feldspar, quartz, and mica. Common at convergent plate boundaries and continental hotspots. Intrusive equivalent: granite. *(L3)*

Richter scale: most commonly used earthquake magnitude scale, based on P-wave amplitudes. *(L3)*

ripple marks: ripples formed in sand by moving water or air. *(L4)*

rock: natural aggregate of minerals. Most rocks consist of more than one type of mineral. *(L1)*

S-wave: second type of seismic wave to be recorded on seismograms (S= Secondary). *(L3)*

sandstone: sedimentary rock composed of sand-sized grains (1/16-2 mm). *(L4)*

schist: metamorphic rock formed from fine-grained sedimentary rocks (mudstones). *(L2)*

scientific notation: short-hand method for writing very large or small numbers in terms of powers of 10. *(L1)*

seafloor spreading: process in which new oceanic lithosphere is formed by volcanic activity at mid-ocean ridges and then displaced progressively outward. *(L3)*

seamount: an isolated submarine volcanic mountain, typically formed above a mantle hot spot. *(L3)*

sediment fan: fan-shaped accumulation of river sediments at the base of a steep slope. If at the base of the continental shelf, termed a *submarine fan*; at the foot of a mountain on land, an *alluvial fan*. *(L4)*

sedimentary rock: rock composed of recycled pieces of earlier rocks, deposited by water, wind or ice. *(L2)*

seismicity: earthquake activity. *(L3)*

shale/mudstone: sedimentary rock composed of clay-sized grains (<1/256 mm); deposited in quiet bodies of water. *(L4)*

shield: core of a continent, where the oldest rocks are found. *(L3)*

shield volcano: broad, gently-sloping volcano built from many thin layers of basaltic lava. *(L1)*

silicic lava: high-viscosity, high silica lava. *(L3)*

sill: igneous intrusion emplaced parallel to rock layering. *(L2)*

siltstone: sedimentary rock composed of silt-sized grains (1/16 -1/256 mm). *(L4)*

slickensides: lines or striae on a fault surface, recording the direction of slip. *(L3)*

solar nebula: cloud of interstellar gas thought to have collapsed under its own gravity to form the Sun and planets. *(L1)*

sorted sediment: sediment in which grain size is uniform. *(L4)*

spreadsheet: computer program for calculation, tabulation and plotting. *(L1)*

stress: unequal forces applied in different directions; causes rocks to deform. *(L3)*

stratovolcano: volcano built of alternating layers of lava and ash, typical of convergent plate boundaries. *(L3)*

stromatolite: finely layered sedimentary rock formed by mats of single-celled organisms, often blue-green algae, that live on tidal flats. First appear in the rock record in the mid-Precambrian. *(L2, L6)*

subduction: return of dense ocean crust to the mantle. *(L3)*

superposition (principle of): principle that in a layered sequence of rocks that accumulated at Earth's surface, the oldest rocks are at the bottom. *(L2)*

supracrustals: rocks formed at the surface of the Earth -- i.e., sedimentary and volcanic rocks in contrast to intrusive igneous rocks and metamorphic rocks. *(L6)*

surface wave: seismic wave that travels only along surfaces between materials of contrasting density. *(L3)*

terrace: flat landsurface adjacent to a river, representing an abandoned floodplain formed at a time when the river flowed at a higher elevation. *(L8)*

topographic map: map that quantitatively illustrates the relief of the landscape. *(L4; see also Appendix IV)*

transform plate boundary: boundary at which plates slide past each other. *(L3)*

travertine: rock composed of calcite precipitated from water at hot springs or in caves. *(L4)*

trench: deep ocean trough at site of **subduction**. *(L3)*

tributary: river or stream flowing into a larger river. *(L4)*

trilobites: early arthropods (like modern crustaceans). Cambrian, Ordovician index fossils. *(L2)*

tsunami: gigantic wave caused by a submarine earthquake. Japanese: "port wave". *(L3)*

tuff: rock formed from compacted volcanic ash. If ash is still hot when it accumulates, tuff may be fused or 'welded'. *(L3)*

unconformity: surface within a rock sequence that represents a period of erosion (or non-deposition); a gap in the rock record. *(L2)*

vein: mineral-filled rock fractures. In some veins, the mineral fillings formed from magma; in others they were precipitated from groundwater. *(L3)*

viscosity: the degree to which a fluid resists flow. *(L3)*

water table: top of the zone in the subsurface where all pore spaces in sediments and rocks are filled with water. *(L8)*

NOTES

NOTES

NOTES

NOTES

NOTES

NOTES

NOTES

NOTES

NOTES